LIFE

posthumanities

CARY WOLFE, Series Editor

(continued on page 244)

LIFE

A MODERN INVENTION

DAVIDE TARIZZO

TRANSLATED BY MARK WILLIAM EPSTEIN

posthumanities 44

MIN
NE
SO
TA

University of Minnesota Press
MINNEAPOLIS
LONDON

The publication of this volume has been supported by a translation grant of the Italian Ministry of Foreign Affairs. Questo volume ha beneficiato di un contributo alla traduzione assegnato dal Ministero degli Affari Esteri italiano.

Published by the University of Minnesota Press
111 Third Avenue South, Suite 290
Minneapolis, MN 55401-2520
http://www.upress.umn.edu

The University of Minnesota is an equal-opportunity educator and employer.

Library of Congress Cataloging-in-Publication Data
Names: Tarizzo, Davide, author.
Title: Life : a modern invention / Davide Tarizzo ; translated by Mark William Epstein.
Description: Minneapolis : University of Minnesota Press, 2017. | Series: Posthumanities ; 44 | Includes bibliographical references and index. |
Identifiers: LCCN 2017009430 (print) | ISBN 978-0-8166-9159-3 (hc) | ISBN 978-0-8166-9162-3 (pb)
Subjects: LCSH: Life.
Classification: LCC BD431 .T14925 2017 (print) | DDC 113/.8—dc23
LC record available at https://lccn.loc.gov/2017009430

I I I I

It might be well worthwhile to track down the origin of the dogma of the sacredness of life. Perhaps, indeed probably, it is relatively recent, the last mistaken attempt of the weakened Western tradition to seek the saint it has lost in cosmological impenetrability.

WALTER BENJAMIN, "CRITIQUE OF VIOLENCE,"

IN *REFLECTIONS: ESSAYS, APHORISMS, AUTOBIOGRAPHICAL WRITINGS*

CONTENTS

INTRODUCTION

A SAVAGE ONTOLOGY

WE, THE MODERNS, LIVE.

But when did we start living? When did modernity begin? When did we become ourselves? Foucault's most intense and dizzying work, *The Order of Things,* follows in the path of these questions. Some of its theses—those that interest most closely here—were the following.

a. Modernity does not begin, as canonical belief would have it, between the sixteenth and seventeenth centuries, with the birth of the new science and the philosophical earthquake provoked by René Descartes, but it begins later, at the end of the Classical Age, more or less with Immanuel Kant. This is an idea that Foucault will never abandon and that he clearly restates toward the end of his life, when he claims that—in the light of his "ontology of actuality"—the rise of modernity and the rise of Enlightenment come to coincide. Foucault's ontology of actuality, or of the "present," is indeed an ontology of modernity (Foucault 1984).
b. But what is modern? Human beings are modern, or better, humanity is modern, provided that we regard it as a historico-epistemic threshold taking the form of a riddle: What are human beings? As Foucault explains, humanity didn't exist before modernity, in the sense that human beings didn't tackle the issue of their own humanity, nor had an "analytic of finitude" yet been developed

in which "man's being is always maintained, in relation to man himself, in a remoteness and a distance that constitute him" (Foucault 2005, 367). Yet, it is thanks to that remoteness that the profile of man as both the subject and the object of his own reflection arises. It is thanks to that remoteness that man appears as an "identity separated from itself by a distance, which, in one sense, is interior to it but, in another, constitutes it" (370). All this delimits the horizon of modernity, or of the "present," whose *critical* threshold is marked, according to Foucault, by Kant's philosophy (Fimiani 1997). It is with Kant that critical man, the man who questions the Self (the Same) of his intimate and critical humanity, sees the light of day. It is with Kant that the anthropological "Fold" surfaces, a Fold by virtue of which human beings will have to first discover, and later adopt for themselves, the law that will finally make them human (Foucault 2005, 372). Ever since Kant, this is what we mean by the *autonomy* of modern individuals. The ontology of modernity is the ontology of autonomy.

c. On the threshold of autonomy, which imposes "the ever-to-be-accomplished unveiling of the Same" (370) or of the human Self, three epistemic historical *a priori* emerge: life, labor, and language. These reorder the entire field of human knowledge: biology, economics, and the sciences of language are born. The presence of biology, a natural science, is quite striking in a work that, from beginning to end, attempts to reconstruct the archaeology and the categorial framework of the human sciences. It is all the more striking when one thinks that ten years later Foucault will see "biopowers" and "biopolitics" as the key to modern power. During the 1970s, Foucault will in fact argue that in the Modern Age a new form of power spreads, one that is detached from the law of the sword, from the classical code of sovereignty, in order to take charge of life as such. Its motto is no longer *Take life or let live* but *Make live and let die* (Foucault 1978). Power, that is, now targets life itself, a life to be cultivated, empowered, directed, and regulated. That said, what are the conditions of possibility of this modern apparatus of power? One, the most important, had been identified by Foucault ten years earlier. It is precisely the emergence of life, together with labor and language, as an epistemic historical *a priori*. Foucault makes the point very clear: before modernity, life did not exist, just as a science of life itself, biology, did not exist. Only with the rise of modernity do a knowledge about life and a power over life appear simultaneously, a biology and a biopolitics whose lethal connections we had better analyze and understand. "Historians want to write histories of biology in the eighteenth century; but they do not realize that biology did not exist then, and that the pattern of knowledge that has been familiar to us for a hundred and fifty years is not valid for a previous period. And that, if biology was unknown, there was a very simple reason for it: that life itself did not exist. All that existed was living beings, which were viewed through a grid of knowledge constituted by *natural history*. [. . .] This, no doubt, is why natural history, in the Classical period, cannot be established as biology. Up to the end of the eighteenth century, in fact, life does not exist: only living beings. [. . .] The naturalist is the man concerned with the structure of the visible world and its denomination according to characters. Not with life" (Foucault 2005, 139, 175–76).

| | | |

That is why we, the moderns, live. Because modernity, the age of autonomy, is also the age of life. We *are* no longer, we *live*: it is a new ontology that sanctions our existential condition, according to Foucault. Knowledge about life (biology) and power over life (biopolitics) are possible because of such a new ontology, such a modern abstraction of life, which is one of the deepest manifestations of our "actuality." But then, what does *life* mean for us? When Foucault talks about it, his prose becomes rather fiery and disordered. We can however gather three suggestions from *The Order of Things*.

Life is a synthetic unity. In the Classical period, the naturalist looked at the living being to name it or to arrange it in the linear space of a representation, of a classification. Only living beings existed, dispersed in their multiplicity, and reduced to a principle of order. The Classical Age is the age of taxonomy. With the advent of modernity, living beings acquire the meaning of "living" no longer by virtue of their relationship to other living beings and their position among them, but rather in relation to the secret and synthetic unity that each of them *hides*. "Multiplicity is apparent, unity is concealed." And from now on living beings will be such "only because they are alive and on the basis of what they conceal" (291).

Life is a secret force. That which living beings hide, that which makes them truly "living," is life itself, understood as a force. The knowledge of classical naturalists was a knowledge about the forms-of-life, whereas the knowledge of those who now look into nature is a knowledge about the force-of-life that animates all living beings and is located in "a sort of focal point of identity," impenetrable to our gaze. Thus, from a *superficial* knowledge about living beings we pass to a *deep* knowledge about life itself. From the knowledge of visible forms we pass to the knowledge of an invisible force. This force, life, is that which bends each living being into itself, it is that which makes of each living being a Self, secretly tending to express and impose its-Self. So, while in the Classical Age "being was posited in the perpetually analyzable space of representation," in the Modern Age "life withdraws into the enigma of a force inaccessible in its essence, apprehendable only in the efforts it makes here and there to manifest and maintain itself" (297).

Life is an obscure will of the Self. At this point we leave the flat ontology of being and representation to enter the space of what Foucault calls a "savage ontology." The field of knowledge is no longer divided into existent and nonexistent things, divided by the border between being and nonbeing, but is invaded by something, i.e., life, that neither is nor is not, but rather *wills to be* within each living being. Each form-of-life is reduced to the precarious and transitory expression of a force-of-life, of a blind and hungry will to exist, of a "silent and invisible violence which in the night devours" all living beings after having brought them into existence. This is the dawn of a new ontology that doesn't know the abyss between being and nonbeing, but dynamically takes up residence in the tension between them. It is the dawn of modernity, of life, that puts an end to the old ontology of the one and the many and replaces it with the savage ontology of the Self. "For life—and this is why it has a radical value in nineteenth-century thought—is at the same time the nucleus of being and of non-being: there is being only because there is life, and in that fundamental movement that dooms them to death, the scattered beings, stable for an instant, are formed, halt, hold life immobile—and in a sense kill it—but are then in turn destroyed by that inexhaustible force. The experience of life is thus posited as the most general law of beings, the revelation of that primitive force on the basis of which they are; it functions as an untamed ontology [*sauvage,* i.e., "savage" in the French original], one trying to express the indissociable being and non-being of all beings. But this ontology discloses not so much what gives beings their foundation as what bears them for an instant towards a precarious form and yet is already secretly sapping them from within in order to destroy them. In relation to life, beings are no more than transitory figures, and the being that they maintain, during the brief period of their existence, is no more than their presumption, their will to survive. And so, for knowledge, the being of things is an illusion, a veil that must be torn aside in order to reveal the mute and invisible violence that is devouring them in the darkness" (303).

| | | |

Now, what "knowledge" is Foucault referring to here? What knowledge unfolds on the horizon of life? What kind of discourse can find a foothold in the context of this "savage ontology"? Although at first one may think of some metaphysics of life and of the many vitalist philosophies that have populated modernity, Foucault is actually convinced that "savage ontology" and its "synthetic notion" of life are the basis of scientific, not only metaphysical, knowledge. He is convinced that the ontological horizon of life includes, or at least contemplates the possibility of, a scientific discourse as well as a metaphysical one.

The scientific side is the one that he most often thinks about and that interests him the most. In fact, according to him, a new form of knowledge is born with savage ontology, a specifically modern type of thought, based on the synthetic notion of life. This knowledge, this new science of modernity, is *biology*—which he places alongside another form of knowledge, *economics*. Foucault is talking about biology when he says that "a system of thought is being formed that is opposed in almost all its terms to the system that was linked to the formation of an economic historicity." If the latter "took as its foundation a triple theory of irreducible needs, the objectivity of labor, and the end of history," biology is instead a system of thought "in which individuality with its forms, limits, and needs, is no more than a precarious moment, doomed to destruction, forming first and last a simple obstacle that must be removed from the path of that annihilation; a system of thought in which the objectivity of things is mere appearance, a chimera of the perceptions, an illusion that must be dissipated and returned to the pure will, without phenomenon, that brought those things into being and maintained them there for an instant; lastly, a system of thought for which the recommencement of life, its incessant resumptions, and its stubbornness, preclude the possibility of imposing a limit of duration upon it" (304). In this passage, we find a further connotation of that synthetic notion of life that, for Foucault, lies at the core of modern biology: life, he contends, is "pure will, without phenomenon." In that regard, we might obviously wonder whether this connotation of life is truly scientific.

The *metaphysical side* is the one that constantly remains in the background, always ready to be born again within the framework of savage ontology. In fact, according to Foucault, the latter contains the seeds of both a science and a metaphysics of life. Biology and vitalism, that is, share the same conditions of possibility. Ergo, the appearance of one entails the (re)appearance of the other. "It is this transition from the taxonomic to the synthetic notion of life which is indicated, in the chronology of ideas and sciences, by the recrudescence, in the early nineteenth century, of vitalist themes. From the archaeological point of view, what is being established at this particular moment is the conditions of possibility of a *biology*" (293). In Foucault's perspective this thesis is central: the archaeology of modern vitalism and that of modern biology are *one and the same archaeology*, since they lead back to the same conception of life, to the same deep semantics of a "savage force." In the final analysis, this is what confirms the soundness of Foucault's archaeological account: the fact that two discourses of a different nature share the same epistemic historical *a priori*—in this case, life. That said, what relationship can be established between a science of life and a metaphysics of life, if both are rooted in the same epistemic historical *a priori*? If for both the secret of life is ultimately "the perpetual devouring of life by life" (302)? Foucault's answer is elusive and takes the form of a further question that avoids the original one and shows the limits of his own investigation.

The answer/question is the following: "Must we admit that from now on each form of [epistemic] positivity will have the 'philosophy' that suits it? Economics, that of a labor stamped with the sign of need, but with the eventual promise of the great reward of time? Biology, that of a life marked by the continuity that forms beings only in order to dissolve them again, and so finds itself emancipated from all the limitations of History?" (304). Foucault never answers this question, either here or elsewhere. One can, however, guess what he is hinting at. Just as economics, understood as a modern science, is always susceptible to sliding into a philosophy of history with which it will share the same epistemic *a priori*, so biology, understood as a modern science, is always susceptible to sliding into a philosophy of life with which it will share the same deep and "savage" semantics. In other words, there seems to be no divide between a science and a metaphysics of labor, just as there seems to be none between a science and a metaphysics of life. But why does Foucault stop here? Why does he not look further into this matter?

| | | |

One of the reasons he doesn't, the most superficial, is that he doesn't need to in the framework of an "archaeology of the human sciences." To conclude his task all he needs to do is underscore the discontinuities between the epistemic paradigms of the Classical and Modern Ages, without providing a detailed map of the reciprocal interactions that are produced within the confines of an identical epistemic paradigm, whether classical or modern, between different regimes of enunciation—for instance, the philosophic and the scientific. An analysis in the archaeological vein is differential and discontinuous; therefore, it doesn't aim to reconstruct epistemic organisms from top to bottom; rather, the purpose of this type of analysis is to highlight gaps and fissures. It should not busy itself with plenitudes but with voids, from whose margins one can discern leaps and fractures. It is true that each epistemic paradigm appears to be traversed by tensors that make its profile more compact here and there. Foucault himself, for example, is perfectly aware that the modern human, haunted by a desire that manifests his "anthropological hollow" (280), does not appear at the same time as modern life *by accident*. This explains some of his most suggestive pages, above all those in which the themes of will and life seem to flow into one another, thus bringing to light a single savage force that comes to coincide with the strange "freedom" of modern humans. A passage like the following is symptomatic because of the manner in which it opens and closes all issues, in the sudden flash of an intuition: "The violence and the endless effort at life, the hidden energy of needs, were all to escape from the mode of being of representation. And representation itself was to be paralleled, limited, circumscribed, mocked perhaps, but in any case regulated from the outside, by the enormous thrust of a freedom, a desire, or a will, posited as the metaphysical converse of consciousness. Something like a will or a force was to arise in the modern experience—constituting it perhaps, but in any case indicating that the Classical age was now over" (227).

There is, however, another reason Foucault decides not to move in the direction that we are trying to point out here. This second reason is, most likely, a justifiable caution. After all, if one arrived to the point of arguing that there is no divide between a science and a metaphysics of labor, everyone knew where one would end up: in the vicinity of the first great metaphysical outpost of modernity, where the souls of Georg W. F. Hegel and Karl Marx rest in peace. And in this case it is quite evident that no one—not even economists—is entitled to say where science begins and metaphysics ends. None could say it yesterday; none, today. But what should we say about the second outpost? To state that between a science and a metaphysics of life there is no divide entails wandering into a minefield. It means confronting one of the most consolidated scientific dogmas and one of the most authoritative names of modern science head on. And in all probability Foucault, in a work that was already quite daring like *The Order of Things,* did not consider it opportune to take things that far.

Whoever has read this book knows what stratagem he finally adopted. He chose not to name the true father of modern biology in order to go and search for the conditions of possibility of such modern knowledge in an only slightly earlier author. Foucault says that it is in Georges Cuvier that the modern synthetic notion of life comes forth, with all its ambiguities and its deep semantics, a notion that goes far beyond the limits of scientific enunciation, straying into "philosophy." It is in Cuvier that life begins to glimmer, behind and beyond living beings, in the constitutive distance of each living being from its-Self. It is in Cuvier, and not, for example, in Jean-Baptiste Lamarck. "Though it is true that Lamarck, by the force of a retrospective illusion, has been overestimated at the expense of Cuvier, though it is true that there is little awareness of the fact that 'life' reached the threshold of its positivity for the first time with the *Leçons d'anatomie comparée,* there is nevertheless at least a diffused consciousness of the fact that Western culture began, from that moment onward, to look at the world of living beings with new eyes" (306).

But why is Cuvier all that important? Because Cuvier, in Foucault's view, takes the step that leads to Charles Darwin. Because Cuvier synthesizes a notion of life that would soon become the heart of the theory of evolution, not only Lamarck's but also Darwin's. "Evolutionism is a biological theory, of which the condition of possibility was a biology without evolution—that of Cuvier" (320). Thus, savage ontology, the ontology that for Foucault is the pulsating ontology of modern life, includes—not only, but *also*—Darwin's evolutionary ontology, it includes—not only, but *also*—the ontology of natural selection, of which Cuvier's biology represents a sort of metonymy (Foucault 1979). And here lies the bridge between a science and a metaphysics of life. Here is the intuition before which Foucault preferred to stop.

I I I I

And yet Foucault knew what we all know. He knew that in modern biology "life should be defined by the possession of those properties which are needed to ensure the evolution by natural selection" (Maynard Smith 1986, 7). Just as he knew that the "doctrine of natural selection" is not just a doctrine that "unifies all of biology" but is the "only possible" doctrine able to fulfill such a task (Gayon 1998, 219). Just as he knew that it was this doctrine that had wedged itself in the gap that had never previously appeared between the living being and its-Self, between "words and things," still perfectly interwoven in the very imaginative naturalist inquiries of the Classical Age (Roger 1963, 53 and *passim*). Just as he knew that it was this doctrine that had made all forms-of-life and all living beings "the result and the temporary halt of a continuous dynamism, which in its turn needs to be defined as 'life'"—life that, so understood, ends up assuming a meaning that invisibly takes it beyond the borders of scientific enunciation (Jonas 2001, 39–40, 53–54). What he probably feared, however, were the misunderstandings and scowling silences such as those provoked a few years later by Karl Popper's famous essay on *Darwinism as a Metaphysical Research Programme.* Popper's thesis was that the theory of natural selection obeys a "situational logic" and is valid only in a specific hypothetical context, in which a preestablished notion of "life" is taken for granted. For Popper, this did not entail denying Darwinism all scientific validity but rather showing its metaphysical edge, which for him remained indistinguishable, for better or worse, from its properly scientific edge. In any case his was not, and was not intended as, a *refutation* of Darwinism (Popper 1976, 194–210). Foucault had been tickled by a similar idea, although circumspectly, in the same period. This is proven, for instance, by a TV debate he held with Noam Chomsky in 1971. Here Foucault argues that in modern biology "life" does not function as any other scientific concept would, but rather as an "epistemological indicator," as a sort of scientific metaconcept, devoid of any real referent and yet fundamental for "designating, delimiting, and situating" biological discourse (Foucault and Chomsky 2006, 6). What did Foucault mean by "epistemological indicator"? How should one interpret a concept that has no straightforward connection to reality but only the role of designating and delimiting an epistemic field? And how should one interpret the last pages of a history of biology published in those years by a well-known scientist—highly praised by Foucault—who dares say that "life" is nothing but a metaphysical entity (Jacob 1993, 315)?

It would be misleading to take these remarks for a general invalidation of all statements and research procedures of modern biology, all to be sucked up into the smoke of a volatile *philosophical theory*. This is not what either Popper or Foucault were thinking of. It is not the case made, for example, by George Lakoff.

> Philosophy is most powerful when it is invisible. Over the course of centuries philosophical theories may become so ingrained in our culture and our intellectual life that we don't even recognize them as theories; they take on the cast of self-evident truth, part of the intellectual landscape that serves as the background for theorizing. Such virtually invisible philosophical theories are often harmless. But when they are false and become widely accepted within important academic disciplines, invisible philosophical theories can stand in the way of scientific investigation. Because they are invisible, they are neither questioned nor taken into account. (Lakoff 1988, 122)

The situation outlined here is almost like a laboratory situation, in which there are watertight compartments, in this case the compartment of philosophical theories, scientifically false and devoid of foundation, and the compartment of scientific theories that are true and well founded. In such a situation—obviously—to demonstrate that a theory is philosophical, instead of scientific, would *ipso facto* be the equivalent of demonstrating that this theory is false and betrays its initial aspirations. But if one looks carefully, this is not the point, because only rarely can we draw clear-cut demarcation lines between registers of enunciation that remain, in many respects, separate. Words continuously emigrate from one region to another of our epistemological horizon and, in doing so, they carry with them invisible meaning-effects. That is not to say that they carry with them entire theories—*consolidated, consistent, unitary*. Enunciations, even if they belong in different discursive regimes, rather appear to contaminate one another, here and there, to blotch one another, obfuscating our futile, albeit legitimate, dreams of purity. In brief, the words we utter, wherever this occurs, are imbued with history and, for this reason, fogged up.

So, it is probably more advisable to see things as Charles Taylor does, in a work that also confronts, very originally, the question of modernity. Taylor's idea is that there might exist, more than invisible philosophical theories, *implicit ontologies,* buried in the "inescapable frameworks," general and widespread, by means of which we orient ourselves in the world (Taylor 1989). Such "inescapable frameworks," Taylor maintains, are historical, they change in the course of time (just like Foucault's epistemic paradigms) and dictate "qualitative distinctions," outlining the confines of alternate moral ontologies. What is philosophy's role in such a perspective? It is not that of inventing and advocating a new invisible ontology, a new moral ontology, almost as if it were a new theory, but rather that, more modestly, of bringing to light, each time, the deep semantics of some nuclear concepts, around which this or that "framework" revolves. The owl of philosophy takes flight at dusk, once the deed is done, as Hegel stated. And yet, it can cast its gaze far away, pointing to the horizon of our world and allowing us to lean out over the edge of its contingency.

| | | |

The savage ontology Foucault speaks of can be considered as an implicit, invisible ontology in Taylor's sense, even though it cannot be reduced to the perimeter of moral values. It is an ontology with strong moral implications, and prior to these, political ones, but it is also an ontology with a profound epistemic fallout. At its center lies a synthetic notion of "life," an abstraction of "life," which is on the one hand the *sine qua non* of biopolitics and even of a religion of "life" that others have questioned (Duden 2002; Dworkin 1993), while on the other it is the *sine qua non* of both a metaphysics of "life," unheard-of before, and a science of "life," which takes place within the same framework. This notion, this abstraction, *does not either exhaust or saturate,* but rather *orients and conditions* the discursivity of "life," in all contexts. Modern "life" has a *density* of its own, wherever we use the word, whether in a philosophical or scientific debate, in a moral or political one. Although, and this needs to be reiterated, such a semantic density does not either reabsorb or recapitulate all discourses about "life" within itself. More correctly, we may say that the word *life* irradiates a secret force that *flexes* all discourses about "life" in certain directions rather than others. And it makes it so that, each time we speak of "life," it is always us, the moderns, who are talking.

| | | |

In this book, following in Foucault's footsteps, I will attempt to decrypt the semantics of modern "life" and widen the frame, which he partially sketched, of savage ontology. In doing so, I will take my cue from a thesis that needs to be clarified. *Autonomy represents the metaphysical threshold of modernity.* It is against this metaphysical background that we can perhaps join what Foucault never managed to join: the semantics of the autonomous will (at the basis, for example, of modern juridical notions) and the semantics of modern life, which are the two sides of our "present." Once we cross the threshold of autonomy, will and life appear as the two complementary faces of our time, as the two sides of modernity that pass one into the other as if they were part of a Möbius strip. But we, the moderns, are neither one side nor the other. We are only the fold that allows the strip of our "present" to wrap around our void. We, the moderns, are not. We live and we will. "The irony of this deployment is in having us believe that our 'liberation' is in the balance" (Foucault 1978, 159). We thus believe that there is a freedom at stake, ours, which should make us human, when inside us a battle rages that has very little of the human to it. "*We do will ourselves.*" [In the original: "Wir wollen uns selbst" (Heidegger 1983, 6)].

1 MODERNITY

THE THRESHOLD OF AUTONOMY

THERE WAS A TIME when free humans crossed the seas, fought wars, debated in the public squares, without laying claim to the will for undertaking such actions. These were the times of Plato and Aristotle, who did not have a word that was equivalent to the Latin *voluntas*. Their world was profoundly different from ours. It was a world where other words were valued: *psyche,* for instance. Human beings were free, souls were free, but not the will. Human beings were even less capable of defining themselves as autonomous individuals. Communities, cities, the *poleis* were autonomous, but not the individual citizens. The first time that the word and the idea of autonomy are applied to a single human being, it is to underscore her tragic, scary exceptionality. That figure is Sophocles' Antigone (*Antigone,* ln. 821), an eccentric character in that world, who was destined to be glorified in quite a different world, ours.

And yet, even though devoid of will, human beings then decided, acted, were motivated by forces whose meaning, distribution, and reciprocal interaction is difficult for us to understand. No doubt these humans lived with *passion.* And for them the diagram of the human being tended to be associated with the diagram of the city and the more inclusive diagram of the cosmos. For Plato, the human's soul was tripartite, just as the structure of the city-state was supposed to be. *Psyche* is thus composed of a rational part (*logistikon*), which aspires to the contemplation of Ideas, and of an irrational and appetitive part (*alogiston te kai epithymetikon*), dominated by our most bestial passions. Between these two levels of the soul there also exists an intermediate one, which Plato calls "the irascible" (*to thymoeides*), and the question he asks is: Is this intermediate part of the soul rational or irrational? His answer is that it can be either. The irascible is, in fact, by its very nature an auxiliary (*epikouros*) of reason, but it is still a passional element of the soul, as is proven by the fact that one can become enraged without reason (*Republic*, 441c). In this sense there exist two kinds of passion in the human soul: on the one hand those that are auxiliaries to reason, on the other those that are clearly irrational, and which reason can only fight, suffocate. The most interesting aspect of Plato's analysis, however, comes immediately after. Plato adds that without the help of the irascible, the rational part of the soul (*logos*) would not manage to bend the appetitive and concupiscent part, the one that lacks any margin of rationality and must be blindly subjected to the orders of reason, to its directions. Reason by itself would not have the *strength* to do so. It would not have the *strength* to counteract the powerful bestial impulses that tend to overcome the soul. But it can find this *strength* in the irascible, which can be moved by reason against the other passions, and is therefore defined as auxiliary to the *logos*. There is no trace of the will in this outline of the soul. Just as there is no trace of it in the outline of the ideal city. The tripartite structure is identical: the counselors or guardians (rational part) are at the head of the city, at its base we find workers and merchants (the appetitive part), and in between there are the auxiliaries or warriors (the irascible part). Once more, without the *strength* of the auxiliaries, the guardians could not govern the community. The irascible's strength supplements reason's purely contemplative asthenia. Within the soul there still does not exist a different force to tame the human animal. The strength of the will does not exist yet.

Rather, a medical-moral schema of the passions seems to rule, which in some passions sees a mere obstacle while in others it sees a prop for the good life, both individual and collective. This is what Aristotle gives us to understand when he talks about friendship (*philia*). On the one hand, friendship is the ethical virtue (*ethike arete*) that both the *Eudemian Ethics* (book VII) and the *Nicomachean Ethics* (books VIII–IX) insist on—"it is a virtue," "necessary to life," which "holds cities together." On the other, friendship is also catalogued as a passion (*pathos*) that the orator must know and manipulate in order to orient the audience one way or the other, in the *Rhetoric*. So virtue and passion are revealed as two sides of the same coin, a coin that bears the imprint of what one could call a *psycho-politics* of human passions. "The ethical virtue, in fact, has both passions [*pathe*] and actions [*praxeis*] as its object" (*Nicomachean Ethics*, 1106b).

| | | |

Such a psycho-political model, which ignores the will, is still influential in a world in which the will instead tends to become crucial. Augustine of Hippo, whom Arendt calls the first philosopher of the will, still bases his analysis of the heavenly and earthly cities on the dichotomy between *amor Dei* and *amor sui* and on the opposition between *caritas* and *cupiditas*, which he describes as the angelic and diabolical faces of human *appetitus*. Thomas Aquinas, some time after, develops a theory of the passions that once more resorts to the Platonic dualism between the concupiscent and the irascible parts of the soul, and once again sees the irascible as a sort of intermediate term between the bestial and divine sides of human beings. But at that point we have already entered a new paradigm, one that detaches humans from the compact bodies of the cities and the cosmos, so as to tie them, directly and individually, to a transcendent God. This new knot, which binds each and every human being, from now on understood as an *individual*, is named *voluntas*. "It is in line with these Augustinian reflections that the Will has sometimes, and not only by Augustine, been considered to be the actualization of the *principium individuationis*" (Arendt 1978, Part II, 6–7).

The vertical relationship between God and individuals, gathered *omnes et singulatim* beneath His gaze, now replaces the circular design of a world eternally enclosed upon itself, without any internal fissures. An abyss is formed between each individual and the next. Each individual is a secret unto the others. And one's secret is the will itself, which makes one that specific individual. Let us take for instance two twins, apparently identical. How can one distinguish them? "The only endowment by which they are distinguished from each other is their will" (109). A characteristic that, while it frees individuals from one another, on the other hand puts them in communication with God on the higher level of a vertical axis, and with their flesh on the lower level. Humans can choose along will's vertical axis, that is the axis of their individuality, whether to stay with God or move away from Him, whether to ascend or descend the steps of the will. The more one rises, the more *voluntas,* i.e., one's individuality, is strengthened, is consolidated, furthering the supreme and sovereign will of God, in which the human will is reflected. The more one descends, the more *voluntas* is weakened, separated from its divine afflatus, thus weakening human beings' individuality and reducing them to animalesque bodies. In this manner, the will, *voluntas,* appears as a *force,* more or less intense, that measures the degree of individuation (and humanization) of the human animal, according to its greater or lesser compliance with the divine will, which from on high inspires and supports all free and authentic human acts of will. *Liberum arbitrium* is born, a type of freedom, a force of will, unknown to the ancients. For them freedom was not an attribute of the will but of action. With Augustine, instead, we witness a "philosophic shift from action to will-power, from freedom as a state of being manifest in action to the *liberum arbitrium*" (Arendt 1961, 163).

Hence a series of questions that only make sense within the ontological horizon of *voluntas.* On the one hand we can ask ourselves up to what point human will is dependent on divine will and to what extent it can detach itself from it, without losing its *strength.* These are the classic paradoxes of divine providence and human predestination: are humans free in the presence of a God whose will they must in any case second, and who nourishes the strength of their will? On the other hand we can ask ourselves to what degree the divine will is enmeshed in the free choices of humans, especially in the evil ones. These are the typical dilemmas of theodicy: is God responsible for the evils we commit? But human *voluntas* is not only enmeshed toward the higher points of the axis. The same occurs at the axis's other end, toward the bottom, where a gradual reconfiguration of the psychic topic of the passions takes hold. Inevitably, once the will has been promoted to human *principium individuationis,* the domain of the passions will also be reconfigured according to the new conceptual geography. And this means that, in the metaphysical world of *voluntas,* desire will become the dominant passion of humans. What is desire, in fact? It is the will's own passion, it is the will mapped onto the plane of the passions, translated and transfigured into passion. If each human being is an individual, and if each individuality is will, the passional trace of this individuality will tend to be desire. This explains the clear insistence, within a similar ontological horizon, on this particular passion, from Augustine's *appetitus* to Descartes's *désir* (up to the Freudian *Wunsch*). According to Descartes, the equation is obvious. Desire is will; to be exact, "the volition to acquire some good or avoid some evil" (Descartes 2015, 29–30). But desire is simultaneously passion, and not just any passion, but rather the queen of passions, the one that pulsates in all the others. So much so that it is precisely by acting upon desire, as if from the inside, that the will shall be able to act on the remaining passions, orienting them and training them with a view to moral conduct. The internal struggle of the will with itself, drawn toward a higher plane by God, and toward a lower one by the weight of its incarnation, thus turns into an internal struggle between will and desire, in the course of which the will is challenged and subjected to continuously renewed trials of strength, whereby it must attest to its freedom. The locus of this passion, of this impact with a resistance that allows the will to express its *strength,* is desire. Morality is the name of its economy. "But because these passions can't lead us to perform any action except through the desire they arouse, this desire is what we should take particular care to control; and that is the main thing that morality is good for" (40, par. 144).

| | | |

Both these conditionings of *voluntas,* at the extremes of the human axis, the high and the low, are part of what Kant will call the "heteronomy" of the will. The human will here obeys something other than itself. Tending toward the higher plane it bends to God's will, who both attracts to him and frees the will of each individual. Tending toward the lower plane it encounters the resistance of desire and more generally of the passions. These are the theological and pathological sides of a will that Kant distances himself from, on the threshold of a further world. This threshold is "autonomy." And it is the *fact* of modernity.

According to Kant, who is the metaphysical engineer of modern freedom, "Autonomy of the will is the property of the will through which it is a law to itself (independently of all properties of the objects of volition)" (Kant 2002a, 58). This means first of all that the autonomous will, unlike the heteronomous will, is no longer founded "on the concept of a self-sufficient perfection (the will of God) as determining cause of our will" and it no longer rests "on a physical or moral feeling" (59). God, who was a principle in the ontological horizon of *voluntas,* now becomes a mere postulate of freedom, of the autonomous will of human beings, that Kant also terms "pure practical reason." The human will no longer depends on the will of God, but exactly the reverse. God is a postulate, an appendix of autonomy. In this sense, the principle of modernity "is yet itself not theological (and hence heteronomy); rather it is autonomy of pure practical reason by itself, because it makes the cognition of God and of his will the basis not of these laws but only of [one's] reaching the highest good under the condition of compliance with these laws" (Kant 2002b, 163). Similarly the passions, once the locus and basis of the will, are now swept away, and with them morality understood as the economy of desires, interests, emotions, and feelings. All this cannot but impair human beings' autonomous will, which in order to be autonomous must be founded exclusively on itself, and must purge itself of any motive that is not the will as such. All else is pathology. "Passions are cancerous sores for pure practical reason, and most of them are incurable because the sick person does not want to be cured and avoids the dominion of the principle by which alone a cure could be effected" (Kant 1996b, 173, par. 81). The only principle that, in this view, must continue to dominate is the abstract will in its pure volition. Kant baptizes it as "the principle of autonomy" and from it derives a "categorical imperative" for human beings. "For thereby it is found that its principle must be a categorical imperative, but this commands neither more nor less than just this autonomy" (Kant 2002a, 58).

The first explicit consequence of this argument is that the autonomous will, the one that is a law unto itself, is detached from all objects or matter, to be reduced to the pure *form* of volition. The will is detached from all concrete and specific volition in order to become the abstract and generic universal of will itself. A universal that is shaped by the empty, hollowed-out form of the categorical imperative of autonomy. The "absolutely good" will, "whose principle must be a categorical imperative, will therefore, undetermined in regards to all objects, contains merely the *form of volition* in general, and indeed as autonomy" (62). The second explicit consequence is that autonomous willing, detached from all objective content, remains tied to the pure *subjectivity* of willing. The will becomes an "I will," the event of a subject who wills and is one with willing as such, with pure practical reason. The will erupts as the very fact of subjectivity and selfhood. "One would have to advance beyond the cognition of objects and to a critique of the subject [*Kritik des Subjekts*], i.e., of pure practical reason" (58). The third explicit consequence, the most surprising, is the one that makes of Kant an insurmountable philosopher of modernity: precisely because it is the pure form and the pure subjectivity of willing, autonomy simultaneously emerges as the pure *fact* of willing, which is the *fact* of modernity itself. If the will subjected to the principle of autonomy no longer rests on God's will understood as a principle (cause) of human volition, nor on physical or moral sentiments understood as a plane for the projection and expression of our will, on what does it rest? On nothing, is Kant's reply. Autonomy is a fact. It is the fact or the event of modern freedom. It is the point at which the ontological assessment of freedom is joined with its historical actuality, with the factual irruption of its own enunciation, which *ipso facto* transforms the principle of autonomy into the principle of modernity, into the principle of a present that appears to itself, detached from the past, and entirely founded on its absolute effectiveness. "So act that the maxim of your will could always hold at the same time as a principle of a universal legislation" (Kant 2002b, 45). This is the *categorical* principle of autonomy. But the only maxim of this universal will is an *actual, effective* "I will," devoid of objects, without any content. This is the reason Kant immediately specifies "The will is thought of as independent of empirical conditions and hence, *qua* pure will [*reiner Wille*], as determined *by the mere form of law*" (45). Such a speculative move, able to synthesize the universal and the effective, the categorical and the actual, without making volition dependent on any content, will prove to be insuperable. If the human will is pure and supported by nothing, if it depends entirely on the subjectivity performing its enunciation, then it ends up coinciding entirely with such an act of enunciation, which becomes the institutive act of an actuality that reveals itself to itself *in the first person*: modernity. This, and nothing else, is the *fact* of freedom. These are Kant's own words. "The consciousness of this basic law may be called a fact [*Factum*] of reason, because one cannot reason it out from antecedent data of reason—e.g., from the consciousness of freedom (for this is not antecedently given to us)—and because, rather, it thrusts upon us on its own as a synthetic *a priori* proposition not based on any intuition, whether pure or empirical" (46).

| | | |

The *abstract form* and the *constitutive subjectivity* of the will are, in these words of Kant, for the first time conjoined in the *bare fact* of autonomy, which is the very event of modernity: a present that stares at itself, perceiving its own actuality. If modernity means to look into the present, to think about our today, then modernity is essentially Kantian (Foucault 1984). It is with Kant moreover that modernity, our present, becomes a pure "thought."

Previously, in the antechamber of modernity, one had certainly insisted on the *constitutive subjectivity* of the will, which lies at the foundation of contractualism and the new legal rationalism, from Thomas Hobbes to Jean-Jacques Rousseau. The latter had even recovered the concept of "autonomy" from antiquity, when it was used to name the political and legislative independence of a city from all the others; by injecting the idea of a subjective will into this concept, Rousseau had unleashed the well-known dilemmas of a "general will," which is and is not the same as the individual will: "obedience to a self-prescribed law is freedom" (Rousseau 2002, 167). Only Kant would then have been able to shed light on such a philosophical conundrum.

Previously, in the antechamber of modernity, one had also attempted to explore the will's *abstract form,* which had for centuries been the restless demon of metaphysics, from Augustine to Duns Scotus and beyond. But Descartes, for example, owed the power of his meditations to a gesture that once again conjoined volition and knowledge, without allowing the will to soar in the formal abstraction of its own willing, cleansed of all matter of volition and thereby cut off from any knowledge. This gesture was the *cogito ergo sum* with which Descartes had sealed both the form of volition and its content in the heightened moment of an experience that made the decision to think coincide with the knowledge one was thinking. In this sharp simultaneity of *willing* and *knowing* "thought," *cogito ergo sum,* what went lost was precisely "thought," absorbed by that experience into the space of knowledge, of representation, on whose surface words and things could at that point mirror one another, without leaving the enigmatic strip of a pure and formal, abstract and unconditioned will hanging. It would have been necessary to start from a different "principle," no longer the *Cogito* but the *Volo,* in order to allow a "thought" (*das Denken*) to emerge that was worthy of this name, in the critical interval between willing and knowing. It would have been necessary to start from an autonomous will, *which does not know what it wills,* in order to separate volition and knowledge once and for all, allowing a "thought" distinct from knowledge to emerge, but one that by now—critically, judiciously—reflects on its own willing. This critical, judicious thought, which categorically separates willing and knowing, making of them two distinct areas of reality, the noumenon and the phenomenon, is the thought of the pure and simple *fact* of willing, of the *fact* of autonomy, of the *fact* of modernity. A thought that protects a secret, articulated in three points, around which the fate of modern humans is doomed to revolve.

The *noumenon*, as the word indicates, is that which one thinks but does not know. That which one knows, Kant explains, are phenomena, which are amenable to objectivation. That which one thinks, and one can only think, is the will, in its autonomous subjectivation. But what does thought think then? It cannot reconnect the *fact* of volition to an object of either knowledge or volition, because this would diminish the will, sapping its autonomy with the heteronomy of motives, known and willed. The will would no longer be a "principle." Consequently in the *fact* of volition one can no longer discern an object of thought, but only a subject. Thought, in other words, thinks itself, it thinks its occurrence, it thinks the event of its own subjectivity, when it thinks the will. Thought thinks the identity of thought and subjectivity, of thought and will, of thought and autonomy, of thought and modernity. And in this thought there is nothing that is really thought (an object), there is only a thinking (a subject), always actual, always active and engaged. To think means to reason, always and everywhere. "[To think for oneself] is the maxim of a reason that is never *passive*. A propensity to a passive reason, and hence to a heteronomy of reason, is called *prejudice*" (Kant 1987, 161).

Reason, from Kant onward, is the name of this thought. Not because Kant invents the concept of reason, but because he provides the philosophical grammar for its revolutionized semantics. Reason thus becomes the identity of thought and will; it becomes "pure reason," which is always, at once, "practical reason." This means that reason (*Vernunft*) is the place where the *strength* of willing and the *strength* of thinking are revealed as one and the same *force*—the power of reason. "Reason is a power [*Vermögen*] of principles and its ultimate demand [for principles] aims at the unconditioned" (283). The *Factum* of reason, the identity of thought and will under the banner of autonomy, therefore appears to itself as the *Factum* of a power, the fact of a strength that is expressed in our judgment (*Urteilskraft*) and requires the unconditioned, freeing us from all conditioning that would *prejudge* our autonomy. The fact of modernity reveals itself in the first person as a force of judgment and an exacting power, which emancipates us from the past. A force, a power that is light. "Liberation from superstition is called *enlightenment*, for although liberation from prejudices generally may also be called enlightenment, still superstition deserves to be called a prejudice preeminently (*sensu eminenti*)" (161). But if it frees us from all previous prejudices, illuminating the present in its actuality and in its separateness from the entire past, where does the power (*Vermögen*) of reason lead us? In what direction does the force (*Kraft*) of judgment push us? Where is the thought (*Denken*) of modernity headed?

The force, the power, the need of reason are its autonomy: that is, its "principle" and, equally, its "end." Yet, one must pay attention to the novel, modern aspect of this end, which is not external but rather internal to reason, and which one cannot know but only think. Reason has no ends outside itself, otherwise it would fall into the prejudicial fault of heteronomy. Reason is an end unto itself. In the identity of a thought that is pure thinking and of a will that is pure willing, reason is not a means but an end. It is the end of its-Self. This is why Kant is careful to distinguish all "relative ends" from the "absolute end." Reason's relative ends are heteronomous; they are the knowable contents of a will that obeys something other than itself, of a will that has not been duly separated from the objects of knowledge and volition, of a will that has not yet been distilled, with judiciousness, with critical attention, into its integral purity. Reason's absolute end is instead reason itself—reasoning for the sake of reasoning, thinking, willing—*which does not know what it's thinking and doesn't know what it wills*, but carries on nonetheless. This is the fact (*Factum*) that makes up our reason, and this is its end. But this end, this purpose is purely formal; it is the form of the purpose, an *a priori* form, a form that precedes any concrete, material purpose. The force of judgment and the power of reason express themselves as an end, as a purpose; they express the purpose of the Self, of autonomy, a purpose that however remains inexpressible, unknowable, as it loses itself in the form, diaphanous and ungraspable, of purposiveness as such. "The ground of this principle is: *Rational nature exists as end in itself* [*Zweck an sich selbst*]" (Kant 2002a, 46). Here the purposiveness of modern reason transpires: not in a relative and concrete purpose but in an absolute and categorical purpose, in an empty final form. While the relative purpose is phenomenal and can be known, the absolute purpose of reason is noumenal and can only be thought. While the first purpose coincides with individual volition, each time conditioned by its own subject matter, the second expresses the unconditioned nature of will as such, abstracted from all individual volitions. While the first purpose is visible, since it belongs to the realm of phenomena, the second is invisible, since it belongs to the realm of noumena, wherein a single "principle" dominates, unleashing the force of thought, the power of reason. This force and this power are "ends in themselves," for they express a purposiveness that is abstracted from any purposes.

Therein lies the secret drama of modernity. If the fact of modernity, that is to say the *Factum* of reason, is a "noumenal," invisible fact, and if it is precisely because of this invisible fact that modernity appears to itself leaving its entire past behind thanks to the intervention of an abstract purposiveness, of a pure *form of purposiveness,* which is reason's noumenal form, this means that modernity appears to itself in the reflection of its own invisibility. Modernity sparks like sheer power, the power of reason, which is an "end in itself," a purpose that as such is never realized but has to be attained for ever and ever. Modernity erupts as a task, a project, divorced from its actuality. And if it does not want to vanish into thin air before even having begun, modernity will have to immediately search for some visibility, so as to render its power (*dynamis*) into an act (*energeia*), its potentiality into an actuality that is fully achieved and finally palpable.

| | | |

The name that, starting with Kant, modernity will assign to its invisible fact is "humanity." It is in humans as humans that modern reason will have to be actualized and find visibility. The *Factum* of modernity thus coincides with the announcement of a *new humanity,* about which one can list three properties.

Humanity is holy. "The human being is indeed unholy enough, but the *humanity* [*Menschheit*] in his person must be holy [*heilig*] to him. In all of creation everything one wants and over which one has any power can also be used *merely as a means;* only the human being, and with him every rational creature, is a *purpose in itself* [*Zweck an sich selbst*]. For by virtue of the autonomy of his freedom he is the subject of the moral law, which is holy" (Kant 2002b, 112). Autonomy is holy, therefore humanity is holy. This means, first and foremost, that modernity is not the definitive end of all values, it is not the realm of the *everything is possible,* nor does it mark the end of metaphysics. Quite to the contrary the noumenon is more alive than ever and traces the axiological design and the set course of modernity. Even though this does not yet endow its invisible fact with visibility.

Humanity is natural. "Now in this world of ours there is only one kind of beings with a causality that is teleological, i.e., directed to purposes, but also so constituted that the law in terms of which these beings must determine their purposes is presented by these very beings as unconditioned and independent of conditions in nature, and yet necessary in itself. That being is man, but man considered as noumenon. Man is the only natural being [*das einzige Naturwesen*] in whom we can nonetheless cognize, as part of his own constitution, a supersensible ability [*Vermögen,* power]" (Kant 1987, 323). Here, finally, humanity becomes visible and with it modernity: in one phenomenal and "natural being," from which a supernatural and noumenal *power* (*Vermögen*) must however transpire, bringing the value and the sanctity of humanity to light. Along these lines, modernity will be able to hunt for its own truth and ultimate achievement.

In fact it is at this point that the two great metaphysical options of modernity are disclosed, the two options of a humanity that is both holy *and* natural, noumenal *and* phenomenal, visible *and* invisible. These are the option of labor and the option of life. Within either one human beings find themselves fighting for their humanity, engaging in a struggle for recognition (Hegel), or in a struggle for survival (Darwin). But why do modern humans find themselves *struggling*? Because of the third metaphysical property of modern humanity. Modern humans find themselves engaged in struggle because they find themselves in immediate intimacy with something that flares up in nature like an untamed and unnatural force: the "human species," a singular "species of beings" (*eine einzige Art Wesen*) that is not actualized in any of its exemplars, but remains in a state of potentiality. Humanity, supernatural, from now on becomes the categorical imperative of each single human being, who is simply natural. Humanity carves itself into human nature as the pure *power* (*potentia*) of humanity, not its *act* (*actus*). And this is why struggle rages on in modern humans. Because humanity is not human beings.

Humanity does not present a human face. "Man can and should be represented in terms of the property of his capacity for freedom, which is wholly supersensible, and so too in terms of his *humanity,* his personality independent of physical attributes (*homo noumenon*), as distinguished from the same subject represented as affected by physical attributes, *man* (*homo phaenomenon*)" (Kant 1991, 65). If human beings don't coincide with humanity, human beings must therefore struggle to find themselves, their Self, their own truth, their humanity, in a process of infinite humanization, which will be given the name of History, social or natural. Within modernity society will become history, nature will become history, and human beings will stand at the center of History as a "fact" in search of full visibility. Being the unrealized and untimely epiphany of humanity, their efforts from now on will be directed toward the healing of their *negative* actuality.

| | | |

The first consequence of human beings' extraneousness to their humanity is that the "principle" of modernity itself, autonomy, tends to evaporate. Kant makes a distinction in this regard, one that will leave its mark on all of modernity, between *positive* and *negative* freedom, a distinction that we could see as coinciding, at least in part, with the one that will be established later between *freedom to* and *freedom from* (Berlin 1969). The first is the private and secret freedom of the *homo noumenon,* while the second is the public and expressed freedom of the *homo phaenomenon.* The first is autonomy, pure will, which only obeys itself: it is the *positive* freedom of willing one's own volition in its formal and foundational abstraction, it is the freedom *to* be one Self. The second is the visible trace of autonomy in the sensible world: it is the *negative* freedom of not distorting one's will, it is the freedom *from* all heteronomous and pathological tendencies. These are the two areas of internal legislation, duty, and of external legislation, law. A clear asymmetry exists between them, because it is the principle of ethical autonomy that, here, lays the foundations for the juridical sovereignty of the law (with no further recourse to divine will). However, what is the outcome of such asymmetry?

It is that the *Factum* of autonomy, the *recto* of positive freedom, only becomes visible in its *verso,* as negative freedom. The freedom *to* be a Self, autonomy, only shines in the freedom *from* everything else, as *non*heteronomy. Again, the actuality of modern freedom shimmers in its nonactuality. The invisible remains invisible and becomes effective only by erasing all its visible traces as it proceeds, thus negating its own appearance. "We know freedom (as it first becomes manifest to us through the moral law) only as a *negative* property in us, namely that of not being *necessitated* to act through any sensible determining grounds. But we cannot present *theoretically* freedom as a *noumenon,* that is, freedom regarded as the capacity of man merely as an intelligence, and show how it can *exercise constraint* upon his sensible choice; we cannot therefore present freedom as a positive property" (Kant 1991, 52). The fundamental distinction between thinking and knowing returns, but so as to now show its diabolical side. If autonomy, the *positive fact* of modernity, is a categorical and positive *saying yes to* one's "humanity" (*homo noumenon*), this fact however only becomes visible in a categorical and negative *saying no to* each "human being" (*homo phaenomenon*). Everything that human beings touch is poisoned; it alters and compromises their pure will, bending it in a spurious and insidious direction. The autonomy imperative is entirely condensed in the inflexible condemnation of any heteronomous impulse. Modernity appears to itself in the sheer *negation* of any past. And it is not only human beings that are buried by such an imperative, but also freedom. For the freedom of the moderns retreats at this point until it is reduced to pure force (*Kraft*), pure power (*Vermögen*), or pure potentiality (*dynamis*), divorced from all actuality (*energeia*). The metaphysics of modernity, the metaphysics of autonomy collapses into a metaphysics of absolute power and potentiality, savage and extraneous to itself, that tends to annihilate freedom itself by reducing it to an untimely, not actual, not active possibility. "The will, which is directed towards nothing beyond the law itself, cannot be called either free or unfree, since it is not directed to actions [. . .] Freedom can never be located in a rational subject's being able to make a choice in opposition to his (lawgiving) reason [. . .] Only freedom in relation to the internal lawgiving of reason is really a capacity [*Vermögen*]" (52).

As a result, modern freedom is no longer conceivable as *liberum arbitrium,* as a freedom of choice between preestablished options, that are subject to its will, since this would not be consistent with the fact of autonomy, which is a pure form of volition, abstracted from all concrete contents. Freedom completely withdraws into itself, like an enigmatic force. The force of the Self. But does anything like that really exist? This question leads us straight to the second consequence of modern human beings' separation from their own humanity, paving the way for a new ontology that is no longer the ontology of the one and the many, but the opaque ontology of the Self. The consequence is that human beings at this point will only be able to *lie*. Is autonomy true? Certainly it is not true in the sense that one can "know" it, represent it, and make it the subject of a science. Yet, it is true in a different sense, in the sense that one can always "think" about it. This is the *Factum* that dictates autonomy. Modernity, or autonomy, always thinks about itself. In this way modernity does not protest the epistemic truth of its thought but rather its testimonial truthfulness. Thought, will, reason, and autonomy are, on "principle," testified to by the Self. *Volo,* we read at the entrance to modernity. The *Factum* of autonomy coincides with its endless thought, with its sheer enunciation. "I will," "I think," "I reason." But is it true? As a matter of fact, if human beings (*homo phaenomenon*) never completely adhere to their humanity (*homo noumenon*), this thought becomes *a priori* mendacious. Thus, from Kant onward, modernity will be obsessed by the impossibility of human sincerity, by the haunting torment of mendacity. A torment that

Descartes, for instance, could not experience: for him doubt was the possibility that God could lie to him, not his *ego*. With Kant instead, and more generally in modernity, doubt shifts, suspicion slides and falls onto human beings. And it is more than doubt or suspicion. "For a man cannot see into the depths of his own heart so as to be quite certain, in even a *single* action, of the purity of his moral intention and the sincerity of his disposition" (196).

Throughout modernity human beings are therefore doomed to lie, also and especially to themselves. For human beings are the face of a faceless humanity, of a faceless Self. A fatally mendacious face. Hence both an internal and an external problem, whose friction will produce the sounds of a new anthropology. The internal problem is the psychic fragmentation of *modern humans*, extraneous to themselves, who will become the object of inquiries all aiming at restoring their lost integrity (*Wo Es war, soll Ich werden,* following the Freudian dictum). The external problem is the political evanescence of *modern humanity,* which will be left without any criterion of visibility and recognition. The abyss between the private freedom of the *homo noumenon* and the public freedom of the *homo phaenomenon* becomes unbridgeable. Private freedom, which is the condition of public freedom, can only be testified to individually. However, there exist no truthfulness criteria that can make this private *Factum,* invisible to others and even to oneself, public. As a result of the bridges between private and public truthfulness having been demolished, it is once again individual sincerity, namely the sincerity and honesty toward oneself, that comes to collide with its very impossibility. There are no public, visible, palpable, traces nor public, concrete, tangible actions from which human beings can unquestioningly infer their private sincerity, authenticity, humanity. Thus, the private and public tragedies of modern humans, in which they furiously thrash about, reveal themselves as the two sides of the same coin. On the one side we read "no public autonomy without private autonomy," on the other we read "no private autonomy without public autonomy." This will be the dissonant refrain of those lame thinkers who are the philosophers of late modern times.

| | | |

Martin Heidegger summed up the tragedy of modernity with concise and cutting words: "The aimlessness, indeed the essential aimlessness of the unconditional will to will, is the completion of the being of will which was incipient in Kant's concept of practical reason as pure will. Pure will wills itself, and as the will is Being" (Heidegger 2003, 101). Nonetheless, the word "being" is not sufficient to express the secret of modernity, nor does it register the depth of its seismic commotion. Indeed, the ontology of modernity revolves around "the nucleus of being and non-being jointly"; it is a savage ontology in which "one is trying to express the indissociable being and non-being of all beings" (Foucault 2005, 303). Within the framework of such an ontology, which we would rather call a metaphysics, it is the Self that comes to the fore, and the Self is a category that differs from being or that radically changes its features, allowing us to avoid the old ontological dilemmas of the one and the many. The metaphysical event, in this new philosophical horizon, is not the naming of being but the naming of an invisible and elusive Self. Throughout modernity, the names that ought to capture it, thus making the invisible visible, will be mainly two: labor and life.

Within the dimension of autonomy, which divides human beings from their humanity and makes humans invisible to themselves, hanging them with the noose of a secret and ineffable freedom, the space that opens up is that of pure "power" (*Vermögen*). *Human beings can be human.* Human beings inhabit their sheer potentiality to be truly themselves. Human beings, accordingly, remain on the path toward their humanity, authenticity, truthfulness. And they must, at this point, build a bridge that takes them from one side of the river to the next, from the invisible to the visible, from the nonactualized to the actualized, from a private freedom, one that is only there *potentially*, to a public freedom, which is also *actualized*. This bridge between human beings and their humanity, between humans and themselves, is History, a bridge on which humans from all over will march with conviction, under the banner of a humanity that is finally free and sincere. They will do it by believing that they can finally see the force of autonomy incarnated, now in a social force, now in a natural force. They will do so by naming their positive freedom *to* (be truly themselves) either as a freedom *to* work, or as a freedom *to* live. This is so even if the metaphysical cage in which they will remain trapped tends to dissolve these two promising freedoms into two rights deprived of any historical directionality: the right to work and the right to life.

| | | |

Humanity must fulfill itself through an infinite process. It is in this formula that the hieroglyphic of modernity is etched. Humanity is excavated "today" within each single human being as a void of Self, as a savage force that devours its profile. In this sense the modern anthropological paradigm is that of a defective human nature. *Human beings are lacking humanity.* Against this background, which we all share, we can take two different paths: one leads us to the past, the other to the future. In the disorderly time of modernity, of an actuality that appears in its lack of actuality, the present is in fact eclipsed, and one can search for its truth, its apparition, its phenomenon, only by looking backward or looking forward. These two alternatives are not equally valuable, though. In the disorderly time of modernity, the most logical alternative remains that which leads us to the future, since our present always appears to itself in a *negative* contrast with everything that belongs to the past. Thus, of the two metaphysical paths, that of the past and that of the future, that of labor and that of life, Hegel's and Darwin's, the former will be the shortest. This is because the metaphysical clock of modernity is moved by an implacable mechanism, and its two hands, namely labor and life, Hegel and Darwin, display the reactionary and the progressive movements of modernity, respectively.

Hegel, precisely by virtue of his reactionary attitude, immediately appears as an outstanding critic of modernity. He immediately grasps its savagery, behind the luminous smile of "reason," and he has no doubts. Modern freedom, autonomous will are *negative power*, forces of destruction. "Only in destroying something does this negative will have a feeling of its own existence. It may well believe that it wills some positive condition, for instance the condition of universal equality or of universal religious life, but it does not in fact will the positive actuality of this condition, for this at once gives rise to some kind of order, a particularization both of institutions and of individuals; but it is precisely through the annihilation of particularity and of objective determination that the self-consciousness of this negative freedom arises. Thus, whatever such freedom believes that it wills can in itself be no more than an abstract representation, and its actualization can only be the fury of destruction" (Hegel 1991, 38). These words, almost two centuries later, still describe the tragedy of "the present," the tragedy of modern human beings, haunted by an abstract and empty will. "This is the freedom of the void, which is raised to the status of an actual shape and passion" (38). But how can one escape from such a void?

The Hegelian reply is so important and symptomatic because it is a titanic attempt to escape modernity by making it finally appear to itself, by making its "phenomenon" glimmer, thus overcoming (*aufheben*) its structural invisibility. (Something that will inevitably detach it from "the present," from that nonactualized actuality of modernity that will find its most natural outcome in a metaphysics that is able to become just as nonapparent and nonactualized: Darwin's *invisible metaphysics*.) Hegel's response to modernity is a "phenomenology" of the human, which on the one hand attempts to dialecticize, and thereby overcome, the modern *negation* of the past, while on the other it attempts to turn the abstract over into the concrete, by perceiving the true humanity and autonomy of human beings in "labor" (*Arbeit*). Hegel's objective is clear: it is a matter of mutual "recognition" (*Anerkennung*) between human beings, who must recognize each other as human, absolutely. The primary scene of this recognition, which marks humanity's entry into History, is the ancient struggle between the slave and the master.

| | | |

To dialecticize the modern negation of the past means to reopen a dialogue, a dialectic precisely, with the past, *starting from its negation*. It means to make of this very same negation the "dialectical" (*das Dialektische*), which in its immediate abstractness will be negative but in its mediated concreteness will become positive. This is what makes Hegel a sort of premodern figure adrift in modernity. It is not so much his attempt to bring Aristotle back into fashion as his attempt to *achieve time*, to *achieve modernity*, by escaping from a metaphysics of pure potentiality (*dynamis*) and reentering into a metaphysics of the completed act (*energeia*). For Hegel to achieve time will mean making its actuality actual, making modernity visible and apparent to itself, overturning its *negation* of the past into a *positioning* of time as such, into a timely, effective (*wirklich*) revelation of History, by means of which not only do humans chase their own humanity, but they reach and incarnate it, stage by stage, figure after figure. If "the present" can be achieved according to Hegel, it is because the historicity of History can be achieved. If modernity can appear to itself and raise itself from potentiality (*dynamis*) to actuality (*energeia*), it is because it can finally appear to itself as historical time, as a present that negates its past *but, by so doing, nonetheless affirms History, and thus achieves, effectuates History.* So that this may occur, as Hegel realizes, it is necessary to move the possibility of History backward and to extend the power (*dynamis*) of a modernity that is now act (*energheia*) to the past, to that past that modernity merely tends to discard. If modernity and History are achieved, it is because it had thus been ordained since time immemorial, it is because the actuality of the present had always been contained in the potentiality of the past. This is the meaning of History's *dialectic*. When does History start, then, that infinite process by which humans accede to their humanity? For Hegel it begins once human beings start to work. That is where the first decisive recognition by humans of their humanity takes place, at the end of the furious struggle for recognition between slave (bondsman) and master. Only the slaves work, while the master enjoys the fruits of their labor. But why did the slave kneel down?

| | | |

In the beginning there was only "motionless tautology: Ego is Ego, I am I" (Hegel 2001, 61). Human beings, or "self-consciousnesses," already existed, but they still did not exist for one another. History had not yet begun. The first step to enter it will be the self-consciousness that awakes from its torpor and realizes it is nothing, it is pure negativity, being a Self only in the *negation* of being an Other, being a subject only "in opposition to the first object." Starting from this moment, self-consciousness, the Self, autonomy (*Selbstständigkeit*) function like the eye of a hurricane, immobile at the center, and moving everything to fall toward its empty identity, its pure selfhood, in which all otherness, all difference from the Self, is devoured. It is from this moment that life starts to roar.

Vital processes in fact are processes in which all that differs from the Self, all Otherness, is constantly negated to be sucked into the empty identity of the "I am I" that, in this manner, unleashes a frantic movement around the Self: the center is immobile, everything around it comes alive and moves toward it, to be swallowed by it. The I, the Self, opens its maw. But the living, that which comes alive and moves around the eye of the hurricane, falling into it, remains external and extraneous to the immobile "I am I." It is for this reason, in Hegel's view, that life *is not* self-consciousness; for him the force of the Self cannot be associated with a force-of-life abstracted away from all living beings, as Schelling thought. With Hegel we still find ourselves within a horizon in which only living beings exist, scattered along the periphery of a self-consciousness that cannot mirror itself in them without negating them and repulsing their eccentricity. The living being emerges precisely in this unbridgeable distance from the Self, "in the opposition between self-consciousness and life," *which turns life into an object, rather than a subject.* An object that the Self invariably wants to consume and absorb into itself, as it cannot tolerate its otherness. The living being thus is an object that continuously becomes the object of other living beings, that eternally consume one another. "[T]he object of immediate desire is something living" (61). In this sense to live means to consume, to eat (*aufzehren*), and it is the Self's centripetal force that instills in living beings the hunger and desire for other living beings, *by rendering them all objects for one another.* Vital processes, in this way, appear as an infinite process of "universal dissolution," like an immense food cycle, which rotates around the motionless and intact center of self-consciousness. "The entire circuit of this activity constitutes Life" (63).

What Hegel describes here with some rapid brushstrokes is an impressive prehistoric scene, since life is not yet History, life is the exact definition of prehistory. Life is not a force that extends itself between human beings and themselves, between humans and humanity, because at this point there does not yet exist a humanity either natural or holy according to Hegel—and the only "force" he speaks of is "consciousness" (46–59), not life. Natural humanity is however the first to appear, and behind it Hegel immediately sees, lying in wait, the profile of a holy humanity, which for him is a historical humanity. "In the case of life, which is the object of desire" (63) self-consciousness encounters an extraneous object, the living being, which it negates and dissolves. This dissolution, although it originates in the center, in self-consciousness, takes place at the periphery, in the mutual consumption of living beings, each the object for the other in the cannibalistic circle of life. Now, Hegel states, not only do living beings eat one another in this prehistoric theater, but some of them also suddenly see one another, they see each other eating one another, they see each other hungry and desiring that food that they consume and simultaneously embody for others. They see one another from the outside, from that ungraspable point that makes each of them stick to its-Self, at the center of the hurricane, and they see one another at once as both victims and executioners, all, within life's food cycle, objects and subjects of the immense massacre, of the "universal dissolution" that moves and destroys them. At that point the script changes. But for what reason? Not because these savage and prehistoric beings agree that it is better to cease hostilities to save their lives, as occurred in the state of nature as narrated by Hobbes. Hegel's vision is much more extreme in its cruelty. The process of humanization, for him, passes through an intensification of the desire of consumption. Once a living being has seen in the other not only an object but also a subject of desire and consumption, it will no longer be sufficient to eat its flesh, that living being will also want to eat the other's soul. It will no longer be sufficient to negate or devour the other as object, that living being will want to negate and devour the other as subject. That living being will want to subjugate the other, and obtain from the other a denial of its own subjectivity, a denial of its own desire, so that the other reduces itself to a thing, recognizing in the one who dominates all the others the only desiring Self, the only "I am I." It is then, in this struggle for the recognition of a single desire, in this struggle for pure prestige, that natural humanity emerges as a "species," as a human "genus" (*Gattung*). It is then that one can begin to speak of an "*I* that is *We,* and a *We* that is *I.*" It is then that History begins to stutter. For it is at that very moment that humanity makes inroads into human beings: through a struggle that will no longer be carried on in the name of life, but rather for the sake of a victory, a crushing and definitive victory, of the Spirit (*Geist*) over life.

| | | |

Consciousness "steps into the spiritual daylight of the present [*in den geistigen Tag der Gegenwart*]" (64). This means that in the first day of History "the present," "today," has already been written, modernity is already accomplished. Its past, which potentially contained it, is life, is nature, but this past, now, in the moment in which "the present" is born, can be *negated,* in order to achieve time, in order to reach "today," in order to turn the sheer potentiality of humanity into its full actuality. Natural Life is the past: a past that is only the potential of the present, the potential of History, an unreal, unactualized potential, which is never truly *past*. Human History is the present: the present of a natural humanity, which from the outset shades into a holy humanity, since from its inception it shades into a historical humanity. But how can one think the passage from Life to History, how can one conceive of the irruption of "the present," of "today," of the time that appears to itself? "The present"—this is Hegel's thesis—should not be thought of as life but rather as labor. Labor is the bridge between human beings and humanity, the channel of their mutual recognition. Labor is the "*I* which is *We,* and *We* which is *I.*"

At the dawn of "the present" day, the master has managed to make another man kneel at his feet. He has extracted the negation of his desire from him, has alienated him from himself, and reduced him to a thing. He has rendered him a slave. The master succeeded by defying death, by negating his attachment to life, by negating himself as a living object, in order to impose himself as a pure desiring Self, hostile to life and dominating all living beings. The slave instead trembled and on the threshold of "independent life [*selbständiges Leben*]" (Hegel 1977, 109, par. 174) he chose life instead of independence, of autonomy, thus precipitating to the other side of the divide, reducing himself to a living object now devoid of desire. The slave has thus detached himself from the Self, has negated himself radically, has turned to the other side, but precisely because he trembled, precisely because he suddenly experienced fear, *not of the master but of his own Self*, precisely because he has withdrawn from the Self and the natural fury of destruction, the slave is the one who, secretly, won the natural struggle between primitive humans. It is the slave who has moved closer to the Self, to pure Self-consciousness and its absolute negation, toward that empty identity toward which everything converges and into which it falls—an identity that is empty and tautological to such an extent that it cannot even wear the mask of the master and of the natural humanity of a desiring subject. It is the slave who has moved closest to the Self, for it is the slave who has moved farthest from it. The Self is in fact a *force* (*Kraft*) that extends itself across the infinite distance between humans and themselves (Hegel 2001, 65), and it is in the slave that this force has shone forth, managing to unfold its play (*Spiel der Kräfte*). It is in the slave that humanity has really shone forth, because in slavery humans have separated themselves from their barely outlined humanity, they have broken the chains of the natural and desiring "human kind," thus coming to light in this departure from their own Self and *ipso facto* opening the path of History, which is a slow and laborious path toward the reconciliation of humans with their own humanity. It is in the slave that the force of the Self has been unleashed with maximum intensity, in its extreme negativity, capable even of negating itself and of instantly quashing natural humanity's desire. In this infinite distance from him-Self, which is by the same token an infinite proximity to him-Self, the slave has remained congealed, in terror, like a cadaver in life. He has surrendered to the "absolute master" of all things, with lost eyes and trembling legs. And it is precisely then that humans appeared on earth, because it is only at that point that humans realized they were nothing, that they were torn from themselves by a Self that, after having given them a natural humanity, immediately asks for its return, in order to fully root them within themselves, in order to install them in true humanity, which is the split humanity of humans, the humanity of History. "For this consciousness [the bondsman's or slave's] was not in peril and fear for this element or that, nor for this or that moment of time, it was afraid for its entire being; it felt the fear of death, the sovereign master [*des absoluten Herren*]. It has been in that experience melted [*aufgelöst*] to its inmost soul, has trembled throughout its every fiber, and all that was fixed and steadfast has quaked within it. This complete perturbation of its entire substance, this absolute dissolution of all its stability into fluent continuity, is, however, the simple, ultimate nature of self-consciousness, absolute negativity, pure self-referent existence, which consequently is involved in this type of consciousness" (68).

This explains why, according to Hegel, the slave reveals the *truth* about the master. The master and the slave are indeed the two faces of humanity, the natural and the historical, blending into one another at the dawn of History. If natural humanity is the humanity of dominators, historical humanity is the humanity of workers. Yet only historical humanity is a holy humanity, which can achieve natural humanity, because it is only in the slave's labor that humanity as such, *homo noumenon,* is split from natural dominators, from natural humans, from *homo phaenomenon,* and enters that process of infinite recovery from the gap between humans and humanity, between humans and their own Self, which is humans' own truth. The ontology of the human therefore becomes a phenomenology of labor (*Arbeit*). Here, labor is understood as the visible manifestation of the—otherwise invisible—force of autonomy: the force of the Self. Labor expresses the phenomenon of the noumenon, namely the concrete actualization of the Self's abstract potentiality: "the present," "today," *the time that appears to itself.* But what does it mean to work? It means to embody the absolute negation of one-Self, as in the case of the slave, and to change that pure *power* that it was, in the unreal and unactualized past of natural life, into an *action* that will be achieved and be real, actual, present. To work means to achieve time, actualize it, give it form and figure. Faced with the reciprocal devouring of living objects, whose forms constantly negate one another in the cruel kaleidoscope that is nature's war, the slave remains dumbfounded, suddenly negating, blocking, halting his voracious impulses. He places himself in the master's service and begins to work to satisfy the master's desire, no longer his own. He fasts. But, by working for the master, he begins to shape and mold natural objects, thereby shaping and molding his very own activity. Thus, the compass needle of "independent life" begins to point to the opposite side: it no longer points to the side of the bare *life* to which the slave had reduced himself, by reducing himself to a slave, but it now points towards the side of independence (*Selbstständigkeit*), of *autonomy,* of the Self—a Self that is sculpted in the shape and form of all the objects he has worked on, which are the shape and form of those who are working them. Along these lines, the slave slowly reappropriates himself, or his Self, by reappropriating life and nature, and by making his product, his artifact out of them, something that the master in his premature victory and self-assured *power* over life and nature can no longer do. The slave instead, in his uncertainty, can only *act* in his service, and now dedicates himself to taming nature, life. He is the one who redeems the past. He is the one who achieves "the present." He is the one who makes time appear to itself. For, at this point, he is the only one who enacts the force of the Self, the force of autonomy, transforming nature and life into the monuments of his own work. The only problem is that, from now on, the slave will not be allowed to stop working even for a moment. In order to always maintain life and nature within the orbit of the hurricane, of the Self, of autonomy, his work will have to be turned into working, his act will have to be turned into an endless acting, which, instant by instant, *halts* life in his works and no longer allows it to live in the natural cycle, but instead in "the present," in the *hic et nunc,* in a History, which are sculpted in each of his artifacts. This is how humans inhabit time and dwell in History. They dwell in a work that is working, in an act that is acting. And day by day they achieve History, that is, modernity. The metaphysical force of autonomy, the secret and ineffable force of thinking and willing, of reasoning for reasoning's sake, is converted into the metaphysical force of working, which translates the noumenal force into a phenomenal force. Potentiality becomes actuality. The force of autonomy becomes labor-force, labor-power. And in the sweaty face of the worker, of the slave, humans finally touch their humanity. "Labor, on the other hand, is desire *restrained and checked,* evanescence delayed and postponed; in other words labor shapes and fashions [*bildet*] the thing. The negative relation to the object passes into the *form* of the object, into something that is *permanent* and remains; because it is just for the laborer that the object has independence. This *negative* mediating agency, this activity *giving shape and form,* is at the same time the *individual existence,* the pure self-existence of that consciousness, which now in the work it does is externalized and passes into the condition of permanence. The consciousness that serves and toils accordingly attains by this means the direct apprehension of that independent being *as its self.* But again, shaping or forming has not only the positive significance that the bondsman becomes thereby aware of himself as factually and objectively *self-existent;* this type of consciousness has also a negative import, in contrast with its first moment, the element of fear. For in shaping the thing it only becomes aware of its own proper negativity, existence on its own account, as an object, through the fact that it cancels [*aufhebt*] the actual *form* confronting it. But this objective negative element is precisely alien, external reality, before which it trembled. Now, however it destroys this extraneous alien negative, affirms and *sets itself up* as a negative in the element of permanence, and thereby becomes *for itself* a self-existent being" (68–69, emphasis added). *Thus humans become humans.* Natural forms, which devoured one another in the eternal cycle of life, are replaced by work and crafted forms, which are monuments of History, and with the force of their labor they bring life into the orbit of the Self, of autonomy, of holy humanity. *Thus humans become workers.*

| | | |

In such a visionary description of humanity's beginnings, as Alexandre Kojève had clearly seen, Marxian and Marxist thought is already contained *in nuce,* with all its claims and its oddities, which often coincide. In Hegel's words we do find the premises of the first great metaphysical option of modernity, that which will look into History through the lens of an economic analysis of social evolution. From such a perspective, History—which is always a link between humans and their humanity—looks like a series of waves that are moved by a social force and display the savage ontology of *labor-power.* Here, labor appears as one of the metaphysical names that modernity has given itself, the secret force of autonomy. According to this view, the slave will have to appropriate his labor, if he really wants to appropriate himself, or his Self, his force, his autonomy, his humanity. But, in actuality, the slave has always already appropriated himself through labor. By working, the slave has always already seized his humanity in the form of labor-power. The problem is that he can reach his humanity only by working, by acting, by actualizing the present, something that he cannot stop doing if he wants to achieve himself and modernity, or History as the time that finally appears to itself. Thus, the invisible becomes visible, the nonactual becomes actual, the past becomes present. But it becomes a ghostly present, which can appropriate its-Self only by adumbrating its own alienation.

It is a present that rams into itself, it is an actuality that is doomed to be endlessly actualized, this is the History of labor-power, echoing a past that it cannot manage to erase once and for all. The *act,* the acting of labor-power, continues in fact to be achieved through the repeated negation of that unreal, unactualized, past that is the sheer *potentiality* of life and nature. A past that the present wants to leave behind itself but cannot forget, for it is precisely labor-power, the act of the present, that always revives the past through its negation. Within such a metaphysical grid modernity, the present, consequently continues to look backward and can only achieve itself in this manner, by eternally detaching itself from a past that it cannot go past. The shape and the form of "the present," the shape and the form of a time that should finally appear to itself and should give shape and form to autonomy, to the *Factum* of modernity, result in a shape, in a form, that incessantly return to origins, reducing History and time *to an absolute present.* History, in a sense, doesn't move. History has always already been achieved. And we are primitive humans. We are not just like them, we are them, we are their wandering ghosts and are destined to remain such so long as we have a present, since our present has already been written in theirs, indelibly.

The History of labor-power was done as soon as it began. Perhaps this is the reason why the moderns have stopped working, have stopped believing in labor as a force for human and social emancipation, have stopped thinking that their autonomy, the freedom *to* be themselves, could be rendered into a freedom *to* work—a curious idea, completely unknown in ancient times, which has fallen into the void of modernity. Once they have finished working, though, the moderns still find themselves faced with two options, in order to believe they are free and, therefore, truly modern. They can withdraw into the force of pure "thought," of "reasoning for reasoning's sake," from which the *power* of modernity is unleashed—in which case, usually, they will be philosophers. Or else, labor having ceased, they can return to life, to that bare natural life that labor couldn't go past, believing that at least in their leisure time they are free: free *to* live, in an eternal Sunday of life that will prove to be more ferocious than even Hegel had feared.

| | | |

If they become philosophers they will have only one preoccupation, that of turning the hands of the metaphysical clock of modernity back to the starting point. That means returning to the tragedy of "the present," which takes place simultaneously on private and public stages. The refrain will always be the same: no public autonomy without private autonomy, no private autonomy without public autonomy. The interpretive keys will remain the same as beforehand: modernity stems from the equation of liberty and reason. The outcome will be predictable: modern identity amounts to an empty *project.*

Richard Hare is one of the philosophers who has taken this path, which leads one back to "reasoning for reasoning's sake." His intent is to provide moral reasoning with a content, realizing the identity of reason and liberty within it (Hare 1965) and finding out the true meaning of the *Factum* of modernity, of Kant's categorical imperative. Providing meaning to the Kantian imperative in such a perspective entails giving the modern project a meaning, so as to avoid its being reduced to a mere project, a naked power (*Vermögen*) or an invisible force (*Kraft*). But can it be done, can one give a content to the force of autonomy? In order to succeed, we would have to give a full and concrete expression to the *Factum* of practical reason, which is the knot of pure thought and pure will. We would have to deduce the positive and actual form of our will from the force of our thought. We would need to issue an effective moral command from it. Let us suppose, then, that we argue in a perfectly Kantian style, that *a moral order that one can universalize is prescriptive and rational.* If we could find even only one command of this type, we would be able to make thought and will coincide in a *Factum* that is reason's practical, tangible, and detectable doing. We would achieve modernity's project. The invisible would become visible. Yet, how can one universalize a command? And what will a moral command that becomes universal and should be valid for everyone order us to do?

To be sure, such a command cannot just rephrase a previously established moral prescription, whose *concrete content* is always tied to the particular (historical, cultural, social . . .) subjectivity of the individual proposing it. If moral reasoning can lead us to the formulation of a universal command, it will have to raise the moral prescription to the dignity of an *abstract form,* whose cogent and ironclad logic will allow one to extend its imperative to everyone, to subjectivity as such, to the diaphanous Self of modernity. The *power* of our thought and the *force* of our judgment will have to raise us above current prejudices and individual subjectivities, so as to allow us to reach the peaks of a universal Subjectivity, thus making the concrete, visible face of every human being shine through into the abstract, invisible face of their humanity: this is modernity's project. And in the final analysis, the problem is not that this magic is unfeasible. The problem is that it will be a black magic trick, not a white magic one. The universal command of reason, in other words, will always take the form of a negative command, not of a positive one, and this command will not fill the gap between human beings and humanity but further deepen it.

Abstract moral reasoning is in fact a form of *a priori* reasoning that cannot dirty its hands with specific moral contents that would prejudice and jeopardize its *a priori* logic. Rather, it must take all the moral, prescriptive, contents, which are already given, and insert them into the moral reasoning machine, which is the only *judge*. This is the machine of Reason itself (the "tribunal of Reason" in Kant's words), whose logical functioning sets up the process of universalization. We thus obtain: on the one side, previously given moral prescriptions, which are the matter of moral reasoning; on the other, logical argument, which is the form of moral reasoning. Form and matter are allergic to any mutual contact, however. If moral reasoning really remains logical and purely formal, as we desire, it will continue to remain *detached* from all the previously established commands that fall into its mechanisms. If moral reasoning instead becomes material, as we also desire, it will allow itself to be tied to one or the other of the previously established commands, which will in turn *vitiate* its purity as well as its logical, formal integrity. In short, the moral reasoning's machinery will flood.

And yet, in spite of all this, such a machine can still produce a universal command, thereby transforming its vice into a virtue. The machine floods and stalls, but it can still turn this stall into a universal command that is always and necessarily valid in the moral field. Magic sounds enticing: *Do not give moral prescriptions to anyone.* This is the highest expression of autonomy; it is a universal command, like the one we required; and it is moreover an imperative that, like the Kantian one, manages to *command* our freedom. That said, what characteristics does this freedom present? It doesn't have the characteristics of a positive *freedom*, of a freedom *to* (be one-Self). It rather seems to have the characteristics of a negative freedom, of a freedom *from* (being Other). The universal command of autonomy, that is, does not prescribe any moral command, quite the opposite. It absolves us from all moral commands. It does not tell us what moral law we must obey; on the contrary it pushes us not to obey and to recuse any specific moral laws. The law of the Self, the law of *autonomy,* just dictates that we do not impose moral laws on others, and therefore dictates that we do not accept the law of the Other, the law of *heteronomy.* As a result, it frees us, it isolates us, it unties us from others, and this is the only way it nails us to ourselves, without telling us who we are. The positive freedom of being one-Self, the fact of modernity, thus instantly returns to invisibility, becoming ungraspable again, dividing human beings from themselves and from humanity, in the "fury of destruction" of a negative freedom that does not pave the way for the advent of humanity, whether it be natural or holy, but makes its profile even more diaphanous and imperceptible. Human beings return to being purely "volitional" beings as usual (198), whose will is empty, also as usual. Again, they appear as those who will "anything" (110), who will to will, who will themselves. Modern humans fall back into the shadows, in that distance from the Self that constitutes them, taking up residence in a "present" that does not belong to them. *Human beings are not, human beings should return to themselves,* and the place for this final encounter of humans with themselves should be humanity. But all bridges between humanity and human beings are broken, unfortunately, so that the enigma of modernity is cast onto the enigma of a "human race" (195) that only exists in its own categorical, albeit ineffectual, announcement.

| | | |

In modern times, we may conclude, each positive freedom *to* tends to be converted into a negative freedom *from* all others, whose catalog makes up the list of our "rights." Our era is the "age of rights" (Bobbio 1996) that sanction and regulate our negative freedom from each other. And all rights, all individual negative freedoms, can be traced back to one single negative freedom, to one single "innate right" from which all other rights stem. "*There is only one innate right,*" Kant proclaims, it is freedom understood as "independence from being constrained by another's choice." This is "the only original right belonging to every man by virtue of his humanity" (Kant 1991, 63). This is the fundamental right around which autonomy revolves. All other rights are just corollaries, detailed specifications that don't alter the invisible framework of modernity by one iota.

This does not mean that one cannot go beyond modernity. Modernity's secret is rather that it always already goes beyond itself on its own. Modernity must achieve itself, actualize itself, and in this drive it proves to be moved by a silent *force* that amounts to its metaphysical impossibility. The theorem of modernity, which is a Kantian theorem, goes as follows: *human beings regularly lie to themselves; human beings cannot but lie to their own humanity.* Human sincerity is impossible. Humans are invisible to themselves. In this impossibility theorem the premises of a time that is desperately looking for its own truth are hidden. Here, to look for its truth means only and exclusively to look for its own truthfulness. Modern humans lie, as Kant often remarks, even though they are not allowed to lie. But why are we not permitted to lie? Because, strictly speaking, we *cannot* lie, or at least we cannot always. If we just lied, if modern humans were a lie and nothing more than a lie, then we would be erased, annihilated by such a lie. We would not only be invisible to ourselves, we would actually cease to be, cease to exist. Being irremediably divided from ourselves, from our humanity, from our Self, we would vanish into the folly of an unsustainable ontological black hole: "I am not." "We are not." This is why the medal of modernity has two sides, which constantly overtake one another, thus completing our metaphysical design. On the one side, human beings lie to themselves, undeniably. On the other, *human beings cannot always lie to themselves; they cannot always lie to their own humanity.*

Caught in this paradox, which makes humans invisible to themselves precisely because they cannot lie to themselves and remain haunted by the inaccessible truth of their own humanity, modern humans then start to oscillate between themselves and their true Self, attempting to establish the *conditions of their own lying* and to articulate a truth that, while remaining extraneous to them, would draw their face reversed, even if only by erasing its traits. Our time is out of joint, as someone bitterly put it. That is to say, "the present" of modernity can appear to itself only in its nonappearance, so that the inaccessible truth of modernity cannot hide in "the present" but must *logically* hide elsewhere, either before or after "the present," either before or after our times, either in our yesterday or in our tomorrow. Our truth, that is, while being unattainable, must hide in our History, which can either look toward the past or look toward the future, depending on where we choose to situate *the conditions of "the present" lies.* These are indeed the two metaphysical options structurally available to modernity, by means of which "the present" can surpass itself, while standing still. The first is the Social History of labor-force, which places the truth of "the present" in the past. The second is the Natural History of the life-force, which places it in the future. In both cases, it is always a matter of *force,* because modernity is a pure force, a pure power, a pure potentiality. Yet, one force is stronger than the other, as it fits the invisible path of modernity better. The History that is destined to prevail in modernity, so long as modernity exists, is the second. It is not the History of labor but the History of life.

| | | |

Natural History is also a history of an "*I* that is *We,* and a *We* that is *I,*" and besides, it is a history that fully respects the most stringent requirements of modernity, since it is integrally and categorically invisible. It is an invisible history first of all because the history of life is not the history of human beings at work, who leave a visible and ghostly trace of their passage; the history of life goes much further back in time, it is a history of nature, whose traces lose themselves in the night of origins about which one can only speculate with eyes wide shut. The truth about this history, its human truth, is placed, nonetheless, entirely in the future, it is not preserved by our past. The truth about natural human beings is not the same as that of social human beings, it is not a truth that was written yesterday, in a time when "the present" always already appears to itself. It is a truth that instead remains to be written, and will remain to be written, so long as life exists and humans exist. It is a truth that therefore perfectly fits modernity's *project.* It is a truth that represents its most *truthful* name, the truth that best speaks its metaphysical extremity. If modernity is an extreme and unachieved project of autonomy, life, not labor, will be the more appropriate name for this extremity, because in the future, not in the past, "the present" can take such lack of achievement to the extreme and make us peek at the last day of humanity, when we will finally reach ourselves, our humanity, at the very instant of our complete disappearance.

Ergo, of the two great metaphysical options of modernity, Social History and Natural History, the economic History of labor and the biological History of life, the first belongs to yesterday, while the second, *de facto* and *de jure,* belongs to tomorrow. Life, and not labor, is our future. We, the moderns, shall live, even after having ceased to work. And since our very own life is for the sake of tomorrow, we do not really "live" today, but currently *survive* ourselves and live through "the present," feeling that life itself, *a true and truthful life,* is coming through in our survival. This does not deny—in fact it confirms—that life, like modernity, remains something that is perfectly invisible. Our life is indeed so invisible that its invisibility, instead of bumping into its ghostly traces, as in the history of labor-force, tends to increase exponentially. Life, which is the last name of modernity, comes to coincide with an invisible savage force, whose invisibility is doubled. Life takes on the sense of an invisible force even before the science of life, biology, starts speaking of it, but then science comes to duplicate this invisibility and to make the primary invisibility of life, its *metaphysical* invisibility, invisible and inapparent to itself. Endowed with such a double invisibility, the metaphysics of modernity can thus mirror itself in the nonappearing metaphysics of life: the metaphysical statement can invisibly reflect itself in the biological statement.

Hence the heterotopia of contemporary philosophical practice. Biology departments are actually the place where the *true* metaphysics of modernity continues to be produced. The privileged epistemological status of a science like biology is a symptom of such a situation. Biology *is the only science of which we sincerely think that "the present" is already the future.* There are no other sciences that enjoy a similar privilege. In this regard, modern biology is truly the mirror of modernity, since its "today" is nothing but its tomorrow. Those who do not think so are not *credible.* They are either not modern or not really *truthful.*

| | | |

It is Kant, once again, who is the first to grasp the course of events. In his old age, in the same year, he publishes two works whose theses complete each other. In the first essay he just rephrases modernity's impasse. Human beings deceive themselves, but this is precisely their worst sin. As usual, Kant insists that humans cannot delve deeply enough into their own hearts to assure themselves of their own sincerity (Kant 1991, 196), and he then concludes that: "The greatest violation of man's duty to himself regarded merely as a moral being (the humanity in his own person) is the contrary to truthfulness, *lying*" (225). It is one side of the metaphysical theorem of modernity: *human beings constantly lie to themselves, to their own humanity.* In his second essay Kant presents the other side, which is just as necessary as the other as a support of modernity: *human beings cannot always lie to themselves, to their own humanity.* But how do they manage not to lie? Kant's answer, in this case also, is a precursor of things to come: human beings will not lie only on condition that they transform their moral strength, the "force of reason," the force of autonomy, into a vital force. "Medical science," as he remarks, is "*philosophical* when the sheer power of man's reason to master his sensuous feelings by a self-imposed principle determines his manner of living" (Kant 1979, 181–83). In a nutshell, medical science becomes philosophical when the bare natural life of humans is seen as expressing the force of their reason, which is the force of a "self-imposed principle," the force of autonomy. All barriers between a natural, vital health and a moral one are removed here. This is not to say that the idea of philosophy as *animi medicina* had not emerged in the past, but the connotation that life now tends to assume is completely new: *life expresses a force, and life is all the healthier the more it is penetrated by the force of autonomy.* This is precisely what a medical-and-philosophical science is concerned with—a moral force that is expressed as a vital force. In this fashion we are expelled from the old medico-moral schema of the passions, and projected outside the premodern context of psychopolitics. Morality and politics will henceforth look directly to life itself, abstracted as a force, with its laws and its precise dynamics, which will set the pace of its invisible history. It is no surprise that Kant does not yet speak of biology. This is due to the fact that the word will only be coined four years later.

2 LIFE

GENESIS OF A METAPHYSICAL PARADIGM

1. The metaphysical *a priori* of modernity reads as follows: freedom is the autonomy of the human will.

1.1. Three possible articulations of autonomy derive from this *a priori*: the law, labor, and life.

1.2. The law could be seen as the reverse side of autonomy. Labor and life, instead, as the obverse of autonomy. Indeed, modern freedom is said in many ways. As negative freedom, it is spelled out by the law. As positive freedom, it takes the name "labor" or the name "life."

1.2.1. In modern times, History has often been understood as a process of positive expression of the force of autonomy. In this perspective, labor and life unfold autonomy's drive by following the trajectories of two alternative vectors that trace the courses of Social History and Natural History, respectively.

1.2.2. Modern law as a negative expression of the force of autonomy, on the contrary, has no History. The law of the moderns can more properly be grasped, at its root, as an ever-changing articulation of one single "innate right," arcane and binding, which in the last analysis is neither natural or social.

2. The metaphysical scheme of modern life makes it the *truthful* expression of the modern will. Natural History is the site of such a weird anamorphosis, which depicts autonomous will within autonomous life. The basic properties of life in the modern world are three. *Life appears as the individuation of an unfathomable Self*: this means that life individuates itself in forms whereby the force of the Self, otherwise amorphous, constantly forges itself. *Life results in a furious struggle for life*: this means that life manifests itself in forms-of-life, but there isn't a form-of-life that can contain the savage force-of-life and can domesticate its intensity once and for all. *Life is kept alive by our will to health*: this means that the living being always tends to seal and to heal the—unbridgeable—gap between its generic force-of-life and its specific form-of-life. If life as such is a force that exceeds every form, a potentiality that exceeds every actuality, then life can live through itself only by consuming all living beings.

2.1. Modern life is the *Factum* of the modern will. I shall try to illustrate the sense of this definition and of the above statements by discussing some canonical texts and developing a philosophical argument from them. The textual analyses will not be exhaustive but are instead meant to selectively illuminate those speculative moves from which there emerges the figure of a world, ours, promised to invisibility.

2.1.1. If the metaphysical question from which Kant's *Critique of Practical Reason* takes its initial cues is "what is the will?" the question at the origins of the *Critique of Judgment* is "what is a purpose?" There is continuity between these two questions. Let's say we ask ourselves what a purpose is. In the end, not only Kant but anyone could reply that a purpose is an object of the will, it is the answer to the question: what do you want? By asking ourselves about the nature of the purpose, we continue, thereby, to ask ourselves about the nature of the will, because there can only be a purpose where there is already a will, a desire, a goal, a need—notions that we can initially all classify as variants of the metaphysical concept of the will.

2.1.1.1. Consider the most equivocal variant, need, and suppose that we ask ourselves if the conduct of a nonhuman entity, of an animal for example, can be directed toward a purpose and can be analyzed in terms of finality or purposiveness. We could first describe the animal's behavior as an action that responds to an organic need and we could then see in the satisfaction of that need a purpose of its behavior, oriented by the animal's will, i.e., by the animal's desire of satisfying its own need. Clearly enough, this would already be a human, all too human, interpretation of animal behavior, it would be an interpretation that superimposes a metaphysical frame on the animal's behavior, the metaphysical frame of the will, which so far has not been either proved or excluded as really pertaining to it. We could therefore interpret its behavior differently. We could perhaps interpret it in a mechanistic fashion (Fichte 2005, 32, 36–41). In brief, it seems that we are forced to waver, in analyzing animal behavior and more generally the behavior of nonhuman living beings, between anthropomorphism and mechanism.

2.1.2. Let's ask once again: what is a purpose? It's an object of the will, as previously stated. It follows that it possesses specific ontological properties. In the first place, a purpose is not a reality but rather a possibility (first property). Unlike an object of knowledge, a purpose does not have to exist some place, it is not something of which we have to affirm or presume that it already exists, but it is something about which we state an as yet unrealized possibility. Second, the possibility that is indexed by the purpose is a possibility that competes with others, and is chosen by the will. In this sense it is the will that simultaneously generates both the possibilities and the purpose. Within the overall field of possibilities the will chooses one, and it is this choice that lets us discern the background of alternate possibilities that are discarded. Ergo, a purpose is one possibility among many and the will traces the circle of possibilities like a compass, with its purpose, its choice, its singular volition standing at the center of the circle (second property). Third, the purpose is always tied to further possibilities. I choose to eat an apple instead of a pear, and all around this possibility, chosen by the will as its purpose, further possibilities emerge—such as going outside, to the fruit vendor, and buying that fruit—which can lead me to the realization of that specific purpose. So on the one hand we have the last and final choice, that of eating the apple, while on the other we have a more or less extensive series of subordinate possibilities, which represent so many intermediate stages toward the realization of the purpose I had set for myself. The entire field of possibilities is plotted by the will, but not all possibilities are placed on the same level. The field of possibilities is divided into means and ends, starting with the primary choice of the purpose (third property).

2.1.3. Recapitulating, not all possibilities are purposes, but each and every purpose is a possibility. The territory of purposes is cut from the territory of possibilities, and both territories' horizon is traced by the will. This entails that the will antedates all possibilities, and that we cannot therefore identify it with any possibility, nor with any choice. Every attempt to assimilate the will to a single act of volition, one that pursues a single purpose, is destined for failure. Adopting Kant's terminology, the pure, autonomous, will cannot be reduced to a spurious, heteronomous, will. The latter presupposes the former; any volition presupposes the existence of the will. This is the crucial metaphysical property of the concept of the will, and Kant's true discovery: "pure will," an autonomous will that is irreducible to a heteronomous will, a will that cannot be reduced to its purpose, a will that precedes and founds the very possibility of a purpose. In this perspective one can claim, as Kant does, that the will is the "form" of the purpose, or the form of a finality that is still devoid of content. But one can also ask oneself how this pure will is able to manifest itself. What will the sensible, phenomenal, expression of a will that is still empty, of a will that is just the "form" of purposiveness, devoid of the materiality of a purpose, be? Where, in the world, can one find traces of an autonomous will, which is a pure will to will, prior to any possibility or purpose? Kant provides his answer in the third *Critique*. The manifestation of the will *as such* in the phenomenal and sensible sphere is the problem he addresses in this work. And as we know, Kant's hypothesis has two sides. On the one hand, the expression of free will, understood as a "form" of finality still devoid of content, is the "beautiful" understood as a "purposiveness without a purpose," or a "purposiveness concerning form," of the work of art (or of a natural object that is deemed "beautiful"). "We do call objects, states of mind, or acts purposive even if their possibility does not necessarily presuppose the presentation of a purpose; we do this merely because we can explain and grasp them only if we assume that they are based on a causality [that operates] according to purposes, i.e., on a will that would have so arranged them in accordance with the presentation of a certain rule. Hence there can be purposiveness without a purpose, insofar as we do not posit the causes of this form in a will, and yet can grasp the explanation of its possibility only by deriving it from a will" (Kant 1987, 220). This is the aspect that Kant calls the "subjective purposiveness" (or "subjective finality") of the object, which he deals with in his *Critique of Aesthetic Judgment*. On the other hand, we find what Kant defines as nature's "objective purposiveness," which he deals with in his *Critique of Teleological Judgment*. This is where we come across the problem of life. Kant states that objective purposiveness can be either external or internal, and this, in turn, allows one to speak of an external "utility" or of an internal "perfection" of the thing that we are judging. The purposiveness of an object, seen as the purpose of something else, is external, and in this case we are dealing with a relationship of "utility" of something for something different. The purposiveness of an object seen as its own purpose is internal, and in this case one is dealing with a relationship of "perfection": the parts and the whole of one and the same thing. Now, this second type of objective purposiveness, i.e., the internal purposiveness, is, according to Kant, typical of living organisms or of those that, at the time, one used to call "organized beings." It is an "*intrinsic natural perfection*, as possessed by those things that are possible only as *natural purposes* and that are hence called organized beings" (254). An "organized being" is therefore a "natural purpose." But what does "organized" signify? As Kant explains, "*an organized product of nature is one in which everything is a purpose and reciprocally also a means*" (255). In other words, an organism is an "end of nature," or a "natural purpose," because, within it, each part exists both as means and end for every other part. One can speak of an organism "when all its parts, through their own causality, produce one another as regards both their form and combination, and that in this way they produce a whole" (252). Here, causality means not only efficient causality but also and above all final causality. In fact, the various parts of the organism entertain a relationship of finality with one another, whose flip side is a relationship of mutual instrumentality. Each part serves every other one and they all serve the organism's organization, they all aim to the production of an organized being. An organism is thus a whole whose unity, integrity, organization is its own final cause. Having said that, Kant specifies that final causes in nature are not knowable because our knowledge is limited to efficient causes, and for this reason we do not actually *know* that "natural purpose" that an organic totality represents; we can only *judge* that we are faced with a "natural purpose," without being able to give any epistemic content to our judgment. What we conclude on the basis of our judgment is that a "natural purpose" exists, an organism exists, but we do not know it, nor do we know its purpose. The internal objective purposiveness of living things means nothing more. It is a purposiveness that, paradoxically, prevents us from knowing the purpose of this or that living being. It is a purposiveness that, accordingly, we could describe as *absolute* and *abstracted* away from all purposes. This kind of "ideal" purposiveness, as Kant argues, makes us slide from a "determining" judgment, which is the source of knowledge, to a "reflective" judgment, which can only reflect the abstract purposiveness of the autonomous will in the abstract purposiveness of the living being. According to Kant, we are only allowed to judge, without ever knowing, that "a thing exists as a natural purpose if it is *both cause and effect of itself*" (249). The organism, the living form, the "end of nature," the "natural purpose," is a final cause whose purpose remains "inscrutable" in our eyes. The one thing one can still add is that, if the organism is both its own cause and effect, then the organism is " a *self-organizing* being" (253). This is what *objectively* makes it a "natural purpose," namely, a being that produces, organizes, and forms itself, and that is thereby endowed with a peculiar "formative force" (*bildende Kraft*). Seen in this light "an organized being is not a mere machine. For a machine has only *motive* force. But an organized being has within it *formative* force, and a formative force that this being imparts to the kinds of matter that lack it (thereby organizing them). This force is therefore a formative force that propagates itself—a force that a mere ability [of one thing] to move [another], i.e., mechanism, cannot explain" (253). In brief, the organism is its own purpose, Kant says, *because it continuously shapes itself,* thus expressing an inscrutable "formative force" that animates it and makes it something alive, something completely different from an inert mechanism. From this point of view, the organism remains alive precisely to the degree that it manifests such a self-shaping force, precisely to the extent that it *continues to form itself and to organize itself.* In this sense Kant's thesis is that the internal objective purposiveness of the organized being shows itself not in a finished and static form but rather in an *endless process of self-formation of the living being,* whose purpose on the epistemic (intellectual) level cannot be known. The finality or purposiveness of the living form exists, but it cannot be cognitively identified. The "intrinsic perfection" of living things is a purposiveness that is abstracted away from all purposes and is paired with an equally abstract *vitality,* which in turn "reflects" the abstract *willfulness* of the autonomous will, of that "pure will" that Kant was the first to discover.

2.1.4. But then, what remains of the organism's "perfection"? Everything remains, so long as one has a good grasp of what "perfection" means in a context like this. If the perfection of the autonomous will is revealed on the moral plane of the human person as the will's infinite perfectibility, and is subject to a categorical imperative of purity and autonomy that twists around itself, on the vital plane of the "organized being," perfection is also transformed into the infinite perfectibility of organization, *which never ceases to organize itself.* This is why, for Kant, the internal perfection of the organism no longer corresponds to the immobile configuration of a static form, but corresponds instead to the mobile configuration of a dynamic form, to the incessant process of organization of the "organized being." From this one derives a—strange but logical—first consequence of this view of the living being. The consequence is that, ultimately, the organism will never be completely organized, will never be finished:

on the contrary, the organism will continuously form and complete itself, *without knowing the purpose of its own activity.* According to Kant, the living being wills itself. This is its purposiveness, its willfulness. It is the pure will of its-Self that makes an organism live. Therefore, just as moral individuals demonstrate their "perfection" only to the extent that they demonstrate their perfectibility, the living being demonstrates its "perfection" only to the extent that it demonstrates its perfectibility, by continuing to organize and shape itself, *by continuing to live.* In this manner the perfection of the living does nothing but "reflect" the perfection of the will. The dynamic is the same and the result is identical: the autonomy of the living in one case, the autonomy of the will in the other. We can, moreover, separate the Kantian concept of the autonomous will into two definitions that implicate one another. On the one hand the autonomous will is a *pure will,* in other words an empty will, devoid of a specific purpose. On the other hand the autonomous will is a *will of its-Self,* in other words a will to will, a will that only wills the will as such. To these two noumenal aspects of the autonomous will there correspond two sensible, phenomenal, empirical reflections of the will. On the one hand we have the "beautiful," understood by Kant as the object's purposiveness without a purpose: this is the expression of the autonomous will in the sense of *pure will,* of a will devoid of purpose. On the other we have the "organism," understood by Kant as both its own means and end: this is the expression of the autonomous will in the sense of the *will of its-Self,* of the will that wills itself. In this latter case the noumenal concept of will ends up "reflecting itself," also in the specifically Kantian sense of the word, in the phenomenon of life. The metaphysics of the will ends up "reflecting itself" in a metaphysics of life, of a life that from now on should be seen as an occult, inscrutable, property of the organism. "We might be closer if we call this inscrutable property of nature an *analogue of life*" (254). To better appreciate the implications of this reflexive passage from will to life, let us go over the "perfection" of organized beings. Kant repeatedly insists on the fact (which is not perfectly consistent with the arguments developed in the first *Critique*) that the *nexus finalis,* the final cause nexus, appears in nature where the *nexus effectivus,* the efficient cause nexus, is not sufficient to grasp the knot of phenomena that appears in front of our eyes (Tarizzo 2004, 39–58). In this case we are faced with a "contingency," something we cannot account for with the intellect (and its categories). Yet reason, since intimate vocation leads it to play this role, tends to account for this eventuality whatever the circumstances. And the only way it can still do so at this point is by inferring the work of a will behind it. "The form of such a thing is, as far as reason is concerned, *contingent* in terms of all empirical laws. [. . .] Hence that very contingency of the thing's form is a basis for regarding the product as if it had come about through a causality that only reason can have. Such a causality would be the ability to act according to purposes (i.e., a will), and in presenting an object as possible only through such an ability we would be presenting it as possible only as a purpose" (Kant 1987, 248). This is the first step in the process of the will's "reflection" in life. The natural organism, the living being, by virtue of its inscrutable contingency in the eyes of reason, bears the traces of a will, of a purposiveness. The second step is no less important. The purposiveness of an organized being is an internal purposiveness; it is not an external purposiveness. The organism is not some other entity's purpose but is its own purpose. This entails that the will whose imprint it bears (and the same goes for purposiveness) is not an external but an internal will. The organism, that is, expresses a will *of its own,* which is precisely a will *of its-Self,* a will to appropriate itself. Which brings us to the third and last step in the process of "reflection" of the concept of will in the phenomenon of life. If the organism, understood as its own purpose, expresses an internal purposiveness, namely a will of its-Self, this means that the organism never ceases, nor can it ever cease, to will itself. This is the secret of the living being, that which raises it to the dignity of a "purpose" (of nature). The organism can only be a purpose (of nature) *in itself* to the extent that *in itself* it tends to the realization *of itself,* to the fulfillment *of itself,* without ever realizing itself, without ever completing itself. This is why Kant speaks of the organism's "formative force," and this is why he perceives an "analogue of life" in this force. Because for him life (organic, organized) becomes the expression of an enigmatic will of its-Self, which reproduces the unfathomable enigma of subjective will in the realm of nature. If the autonomous will of the Kantian person is a will that withdraws from all empirical, phenomenal, sensible content, to reduce itself to pure will, to a will of its-Self, to an anonymous will to will, organic life is nothing but the natural expression of this same will, is nothing but the never-ending "reflection" of this will in nature. From this, two further consequences follow: first, at this point organic life becomes a life imbued with subjectivity, with aseity, a life that is always searching for itself; second, this search moves along the blind alley of a pure will of its-Self, devoid of content, of a pure will of life, devoid of any specifications. Life surfaces in nature like a "formative force" that, precisely as such, precisely in that it is formative, does not possess any form *in itself,* but exists prior to any form, nourishing the process of continuous *formation* (or continuous organization) of living beings. Life retreats to a position prior to any natural form, thus coming to coincide with the ceaseless process of formation of natural forms. Life becomes bare life, without any qualification, since life is now the source of all qualifications. And from here it will be only a short step to arrive at two bizarre results, albeit logical ones, of this purely metaphysical (not yet scientific) articulation of the concept of life. On the one hand, life will become identified with the mere *survival* of the living. On the other, life will be identified with an obtuse *will to health* of the living. If in fact one perceives in the life of living things an anonymous will to will, which does not will anything specific, the purpose pursued by this will shall also be indeterminate, it will be a life reduced to the bare *survival* of the living. Which, in its turn, will allow one to observe an inbred, insatiable *will to health*—an obscure yearning for the healing of one's own, always perfectible, (mal)formations—gush forth from the living. Life as will to life, will to health, will of living things to heal, tirelessly chasing a form, a qualification, an organization, of their own. It is in this abstract and imperative life that the abstract and imperative will of Kantian subjectivity is "reflected." Life, thus, appears as an encrypted manifestation of human freedom. This is the lesson the older Kant imparts. The organism's good health shall, then, be converted into tangible proof of the person's good will.

Autonomous life shall, to all effects and purposes, become the sensible, empirical, phenomenal representation of the invisible autonomous will, or the only concrete *value* that the human will can still express. Hence the—apparently abstruse—words written by a now-aging Kant, with a trembling hand: "Old age therefore claims to be considered something *meritorius*" (Kant 1979, 179, emphasis added).

2.2. We here find ourselves at the origins of what we could define as the defective paradigm of human life and of life as such, which still seems to prevail in reflections of a philosophical (and sometimes scientific) character on the topic. We are not dealing with an epistemic but rather a metaphysical paradigm, which embraces and circumscribes an entire horizon of speculations, whose voices, for this very reason, belong to a single world. We, the moderns, live. *Our life is the expression of a pure will, of a will to will, which finds its crucial manifestation in life.* How can one reconnect the autonomous will, detached from all heteronomous impurities, to the sensible world? Life is our answer to this question. Life is the anchor that secures the metaphysical, noumenal, will to the empirical, phenomenal, world. If the Kantian person's pure will condemns us to the pains of a negative freedom that refuses all sensory inclinations, turning his back to the world, since the dawn of modernity life surfaces as the expression of a residual positive freedom, of a will that has to attest and embody itself in the realm of phenomena so as to will itself without immediately vanishing into thin air. *Sozein ta phainomena,* this has always been the task of metaphysics. Saving phenomena by saving itself, this is the task of the will to will, transformed into an obscure will to life. But, in such a perspective, to *save* shall only mean to *heal,* by satisfying a dark, categorical, imperative to life. It is thus the metaphysical angle that rotates one hundred and eighty degrees. With Kant we bid adieu to all transcendent values and principles that an older heteronomous will could still hold on to. With the discovery of the autonomous will, which is the factum of modernity, we, the moderns, pass from the reign of transcendence to the reign of pure immanence. From a world centered on God and the issue of *salvation* we penetrate into a different world, dominated by human beings and the issue of *health*. In this world it is no longer a matter of looking questioningly to God, with fear and trembling; instead, we must look ourselves in the face, to ask ourselves what our life is: a life detached from every otherworldly crutch, a life all curled in on itself. This is the point at which a metaphysics of autonomous life arises from the incandescent embers of a metaphysics of autonomous will. And this is also the moment in which modern anthropology mumbles its first words. What are living things? What are human beings? In a world that revolves around the will to will *qua* will to life, the answer to these two questions will be both univocal and surprising: if living things are those that by definition chase their own vitality, human beings are those that by definition chase their own humanity. *Was ist der Mensch?* Each time, the question will bypass the answer and reemerge, hard as a rock.

2.2.1. When reconstructing the genesis of modern life, it is best not to be blinded by dates. Let us take Johann G. Herder, a pupil and contemporary of Kant. It is sometimes argued that in him we can find the germs of a defective conception of human nature. Now, while it is true that one can find elements in Herder that appear to go in this direction, it is just as true that in his work we cannot find that theoretical juncture of life and will that constitutes the entryway to modern metaphysics. In other words, for Herder human beings are not constitutionally defective, sick, and incomplete. On the contrary, human beings are perfect, as, even though to a lesser degree, are all other remaining creatures (Schlanger 1971, 140–43; Dumont 1986, 137, 228). One should not forget this when perusing his observations on human beings' "softened instincts," where one is tempted to catch a glimpse of the modern anthropological paradigm. If according to Herder it is true that man, the "teachable creature," "must learn, as he receives from Nature less knowledge: he must exercise his powers, because he receives less power from Nature" (Herder 1966, 62), it is also true that elsewhere he writes: "Men repeat after one another, that man is void of instinct, and that this is the distinguishing character of the species: but he has every instinct, that any of the animals around him possess; only, in conformity to his organization, he has them softened down to a more delicate proportion. [. . .] Man therefore is not properly deprived of instincts; but they are *repressed* in him and made *subordinate* to the dominion of the nerves and finer senses. Without them the creature, who is still in great measure an animal, could not live" (89–90). From these words one can easily gather that Herder does not believe that human beings exhibit an instinctual lack. On the contrary, he criticizes as "foolish" those who lament a presumed "human weakness" (90). It is not that human beings possess fewer instincts than other animals; the point is that in them instincts are attenuated, that is, repressed and organized. Which does not disprove, in fact it rather demonstrates, that "man seems to be among animals that excellent middle creature, in whom the most numerous and subtle rays of similar forms are collected, as far as consists with the peculiarity of his destination" (40). But why are human beings, among all creatures, the "excellent middle" ones? Because, actually, human beings possess the most complex organism—Herder defines it as a "prototypical" organism—in which the organic and instinctive traits, which are divided and apportioned in animals belonging to all other species, are mirrored and merged all at once. As he writes, this depends *"on the more perfect organization in the whole"* (79), which relieves the bonds of blind natural necessity, and elevates human beings to the ranks of a superior creature, both rational and endowed with free will. Human beings, in reality, do not possess *fewer but rather more* instincts than other animals do, and it is precisely for this reason that more paths open up for them so that they can freely choose between different alternatives. "As every beast remains true upon the whole to the qualities of his kind, and we alone have made a divinity of will, not of necessity" (67). Herder's philosophy does not therefore belong to that metaphysics of human defectiveness that will perceive a lack, a void, a defect in human beings—a lack that amounts to their very *humanity*. Nevertheless, the fact remains that Herder's ideas are framed within a philosophy (of the history) of humanity. So, in what sense does he speak of "humanity" (*Humanität*)? Certainly not in the same sense as later, post-Kantian, metaphysics will do. In some respects, one could even state that Herder only pretends to ask himself the question: what are human beings? Because the answer he gives, following a tradition thousands of years old, seems to point to a different query: what is God? "I wish I could extend the signification of the word *humanity,* so as to comprise in it every thing I have thus far said on the noble conformation of man to reason and liberty, to finer senses and appetites, to the most delicate yet strong health, to the population and rule of the Earth: for man has not a more dignified word for his destination, than what expresses himself, in whom the image of the creator lives imprinted as visibly as it can be here" (98). Here the humanity of human beings tends to coincide, in every aspect, with their divinity. That is the reason, Herder emphatically declares, "*religion* is the highest humanity of mankind" (103). Against this background, one can tell in what sense his fleeting observations on life as "fracture and wound" or on the "always fragile and often false health of humans" (123, 124) should be interpreted. In a perspective like his, the problem is not that we are affected by natural diseases and organic insufficiencies. The problem is sin, our spiritual insufficiency. Since worldly life is no more than "a field for practice, a place for preparation" (124) for true life and Humanity, which are an otherworldly life and Humanity, Herder is thinking about *another world, a transcendent world,* when he writes that "the good creator has concealed from us the form of that world, that our weak brain might not be dazzled, or a spurious premature desire excited in us. If with this we contemplate the progress of Nature in the species beneath us, and observe how the artist rejects the more ignoble, and mitigates the claims of necessity, step by step; while on the other hand she improves the spiritual, improves the refined, and animates the beautiful with superior beauty: we may with confidence trust the invisible operating hand, that the *flower of our bud of humanity* will certainly appear, in a future state of existence, in a form truly that of *godlike man,* which no earthly sense can imagine in all its grandeur and beauty" (125). The entire universe is thus outlined as a creature tending toward its Creator, a creature whose head is represented by human beings, who are staring at eternity. The "life" of this universe is not at all a life that is famished for life, a life that tends to grow and wrap around itself, but is rather the sum of the organic "forces" that are intrinsic to every organic "form"—*a stable and invariant form.* The game of life, that is, is not an open game but a zero-sum game, in which everything has been preestablished in detail the day of the creation. The game of life is not a dynamic game, the game of a mobile life detached from all forms-of-life, but the static game of a life nailed to the forms-of-life by a divine hand. For Herder, accordingly, life only describes a vast field of forces, displayed in a hierarchical and ascending line, a field of natural "powers," inserted in so many natural "forms," created and preordained by God so as to analogically allude to another world, to which human beings are predestined. A thesis in which one does not hear the critical announcement of modern life but instead the ancient and stentorian echo of an infallible transcendence. "When the door of creation was shut, the forms of organization already chosen remained as appointed ways and gates, by which the inferior powers might in future raise and improve themselves, within the limits of nature. New forms arise no more: but our powers are continually varying in their progress through those that exist, and what is termed organization is properly nothing more than *their conductor to a higher state.* [. . .] What the all-vivifying calls into life, lives, whatever acts, acts eternally in his eternal whole [. . .] It was not our reason that fashioned the body, but the finger of God, organic powers" (109, 112, 114). To conclude, in Herder's universe there is no trace of a life immanent to itself, for his universe is dominated by the transcendent verticality of the divine. There is no autonomous life opening up natural paths that are irreducible to a predetermined model of development. In Herder's world the vital *process* is still caged in a vital *order,* and *immanence* is occluded by *transcendence,* because life is not permeated by an autonomous will, but rather follows a heteronomous will: it does not so much obey its-Self as God. That said, aside from Herder, where can we find the first traces of the modern metaphysical paradigm, that defective paradigm of autonomous life that Kant begins to sketch? If the hypothesis of a will–life nexus is correct, we should see the first traces emerge in those thinkers who take their cue directly from Kant, trying to follow in his footsteps. And it is indeed in

Johann G. Fichte and Friedrich W. J. Schelling that a defective conception of human life and of life as such first make headway.

2.2.2. "Critical philosophy is thus *immanent*, since it posits everything in the self; dogmatism is *transcendent*, since it goes on beyond the self" (Fichte 1982, 117). This typically Kantian-sounding sentence of Fichtean metaphysics can be read in the *Foundations of the Entire Science of Knowledge*, which is the first edition of the treatise Fichte will then work on for the rest of his life. Curiously, however, the true foundations of the entire science of knowledge are not laid out in the work that bears the same title—even though this is the direction the first revision of the *Science of Knowledge* already moves toward (Fichte 1959)—but instead in a text that is published three years later, *The System of Ethics*: "Nothing is purely true but my self-sufficiency" (Fichte 2005, 17). Here the term used to express the idea of "autonomy" is self-sufficiency, *Selbstständigkeit*, and from now on we will take it for granted, as is the custom among interpreters of Fichte, that this term is a synonym of Kant's *Autonomie* (Cesa 1994, 103). Fichte's idea, to put it in his own words, is that for the problem "To think of oneself, merely as oneself, i.e., separated from everything that is not ourselves," there is but one solution: "I find myself as myself only as willing" (Fichte 2005, 24). Such is the basic assumption of his metaphysics of I-ness (*Ichheit*), which is a synonym of subjectivity (*Subjektivität*) and of aseity (Fichte 2000, 18). But then the issue that had already arisen for Kant immediately reappears in Fichte, leaving a mark on his entire philosophy, or at least on its stellar beginnings. If "the essential character of the I, through which it distinguishes itself from everything outside of it, consists in a tendency to self-activity for self-activity's sake; and this tendency is what is thought when the I is thought of in and for itself, without any relation to something outside it" (Fichte 2005, 34), in what fashion shall this "essential character" of the I, that is, its autonomous will, its willing for the sake of willing, its spontaneous activity for spontaneous activity's sake, manifest itself in the external, sensible, phenomenal world? This is the problem that underlies all of Fichtean metaphysics, a metaphysics of relationship and "mutual exchange" (*Wechsel*) between the I and the not-I. Without entering into excessive detail, let's examine its central tenets.

2.2.2.1. Fichte starts from Kant's definition of subjectivity in terms of pure will and autonomy. Subjectivity (or I-ness) is the activity of being free (or rational) that infinitely "returns" upon itself (Fichte 2000, 18). Subjectivity thus describes the circle of the will to will, which continuously finds and searches for itself, but finds its-Self only by searching and chasing its-Self ad infinitum. In this manner, the will detaches itself from any sensible content, and from any worldly purpose, in order to return to itself as the absolute purpose of its own will. "The sole determining ground of the matter of our action is [the purpose of] ridding ourselves of our dependence upon nature regardless of the fact that the independence that is thereby demanded is never achieved. The pure drive aims at absolute independence; an action is suitable to the pure drive if it is also directed toward absolute independence, i.e., if it lies *in a series [of actions], through the continuation of which the I would have to become independent.* According to the proof that has been provided, however, the I can never become independent so long as it is supposed to be an I. Consequently the final end of a rational being necessarily lies in infinity; it is certainly not an end that can ever be achieved, but it is one to which a rational being, in consequence of its spiritual nature, is supposed to draw ceaselessly nearer and nearer. [. . .] My purpose lies in infinity because my dependence is infinite. [. . .] I am supposed to be a self-sufficient I; this is *my* final end" (Fichte 2005, 142–43, 201). The reason why the I's "pure drive," which is a drive aiming at autonomy and independence, can never be fully satisfied is that the *I itself is such a drive (Trieb).* Said in a slightly different way, the I is not a substance but a process of continuous appropriation of oneself or of the Self, which runs parallel to a continuous process of self-emancipation or of emancipation of the Self from nature and the sensible world. It is for this reason that the "pure drive" is also defined by Fichte as a "mere negation" (of any sensible content), a negation of which the I is not even conscious. "For it has just been shown that a drive, insofar as it is a *pure* drive, a drive directed toward a mere negation, cannot appear within consciousness at all. There is no consciousness of negation anyway, because it is nothing. [. . .] The pure drive is something that lies outside of all consciousness; it is nothing but a transcendental explanatory ground of something in consciousness" (144). Ultimately, the pure drive is the unceasing negation of the Other (of the not-I, of heteronomy), by means of which the Same (the I, autonomy) just as ceaselessly affirms itself. And one is never conscious of that drive, Fichte explains, because that drive constitutes and sustains consciousness, I-ness, subjectivity, which is always about to return to itself. What does this mean? First of all it means that subjectivity, pure drive or pure will, only affirms itself by continuously detaching and abstracting itself from a surface of objectivity, the surface of the natural and sensible world, which the I determines and shapes *through continuous negation.* This is why, according to Fichte, something is raised before the I that we can only define as the *not*-I. And this is why there is constantly a *Wechsel* between the I and the not-I, there is "reciprocal exchange" and reciprocal determination between them. Because, in the final analysis, the *not* of the not-I, is nothing but the I itself. "As we can also put it therefore: the ultimate ground of all consciousness is an interaction [*Wechselwirkung*] of the self with itself, by way of a not self [i.e., a not-I] that has to be regarded from different points of view. This is the circle from which the finite spirit cannot escape, and cannot wish to escape, unless it is to disown reason and demand its own annihilation" (Fichte 1982, 248). Thus, the not-I, or the not-Self, emerges as the photographic negative of the I, as Fichte tries to illustrate. But let us now focus on a concrete example of *Wechsel* between the I and the not-I. Consider human beings, taken as sensible and natural entities. Following in the same path as Kant, after having stated that on a certain level (pure, noumenal) "I find myself as myself only as willing," Fichte specifies that on another level (empirical, phenomenal) "I find myself to be an organized product of nature" (Fichte 2005, 117). Logically, if this is the case, then there must be *Wechsel* between the two levels, which in Kant only tended to "reflect" themselves one in the other. So, in what way shall the (human) will be "changed" into, or "exchanged" for, (human) life? Fichte's answer consists in a three-step argument, based on one single principle: "The I is the first principle of all movement and of all life, of every deed and occurrence" (90).

2.2.2.1.1. First step. One must distinguish freedom, the autonomous will or absolute spontaneity, from "bare life," just as we distinguish the I from the not-I: this is perhaps the first occurrence, at least the first that is pregnant with philosophical significance, of the expression "bare life" or "mere life" (*bloßes Leben*) in a modern metaphysical text (Fichte 1982, 262).

2.2.2.1.2. Second step. One needs to understand in what manner bare life is "changed" into absolute spontaneity, by weaving together phenomenological observation and metaphysical speculation. In this case, too, Fichte does no more than utilize and reformulate several mainstays of Kantian practical philosophy, deepening their ontological implications while using a partially new lexicon. The basic idea reads as follows: the I must be "changed" into the not-I, my spirituality must be "exchanged" for naturalness, without the coincidence of the two specular sides of my being vanishing in the course of this "change" or "exchange" (*Wechsel*). "Are my drive as a natural being and my tendency as a pure spirit two different drives? No, from the transcendental point of view the two are one and the same original drive [*Urtrieb*], which constitutes my being, simply viewed from two different sides. That is to say, I am a subject-object, and my true being consists in the identity and indivisibility of the two" (Fichte 2005, 124–25). Hence Fichte's distinction between a pure, spiritual, drive and one that is impure, natural; both are rooted in a prime, originary drive (*Urtrieb*), which is decomposed into its two complementary aspects. The natural drive is described as a "drive towards self-preservation," which matches a drive towards formation or organization. Once again, we find here the Kantian idea of a "formative power of nature" (116). In brief, "a product of nature that no longer organizes itself ceases to be an organized product of nature; for the character of what is organized consists in its continuing formation" (117). And yet, Fichte argues, it is only in human beings that the natural drive toward organization (of oneself) is combined with spiritual reflection (about oneself). It is only in human beings, therefore, that one witnesses a *Wechsel* between the spiritual drive (toward absolute spontaneity) and the natural drive (toward organic formation). Only human beings, who reflect on their natural drives, manage to transform them into spiritual drives, bringing both drives' common roots into the open. And "the first thing that arises from reflection upon this drive is a *longing*—the feeling of a need with which one is not oneself acquainted" (119). Thus, Fichte goes on to explain, the "longing" (*Sehnen*) emerges as the gray zone, the transitional area between spirit and nature, between the I and the not-I. In this longing, according to him, we can discern human beings' keystone, in which both spirit and nature end up coinciding. In the longing, in the feeling of an undecipherable need, the spiritual drive, that is pure negativity, is in fact "changed" into a natural drive that also bears the traces of an obscure negativity, of a pure defectiveness, of that *not* lying at the core of the I. The longing is indeed the feeling that "*we are missing something; we don't know what*" (119). And this is precisely the indeterminacy point between the I and the not-I, between spirit and nature, between will and life, we were looking for. In this point the two apparently opposite aspects of Fichtean ontology flow back into their originary unity. The longing is not *per se* either spirit or nature, will or life, not because it is not either one or the other but because it is both simultaneously. "Nature does not produce a will; nor, strictly speaking, can it produce any longing, for this too, as we have already seen above, presupposes an act of reflection. In the sort of reflection that is involved in longing, however, the I does not become conscious of itself as engaged in this act of reflecting; and thus the I itself must assume that the longing that is present in it is an act of nature" (150). A spirit that reflects on itself without consciously doing so, a nature that presupposes an unconscious reflection on itself, the longing is seen by Fichte as being the crucial linchpin of the doctrine of science, because it is only thanks to it that the I establishes itself as the *not* of the not-I, and it is only thanks to it that the will takes up residence *as a void* in the heart of life itself, allowing the *Wechsel* between nature and spirit to occur. "This longing is of importance, not only for the practical, but for the entire Science of Knowledge. Only thereby is the self *in itself*—driven [*getrieben*] *out of itself*; only thereby is an *external world* revealed *within it*" (Fichte 1982, 266). That said, if the longing is the imprint of spirit in nature, or the imprint of will in life, it is so because it is the print left in natural human beings by a negativity that defines the very subjectivity of the I, which Fichte (following Kant) conceives of as a pure will of oneself. As a result, there arises (from the incandescent embers of Kant's metaphysics) the figure of a new human being, whose natural existence is as if uncompleted by something that distorts its image and consistency. A void identical to that which agitates spiritual man is now dug within natural man, driven by a blind will of himself. The longing disturbs natural man and throws him off balance, eliciting an uncontainable drive toward self-satisfaction, toward the satisfaction of a silent and unknown desire (*Trieb*). Human life is thus finally riven by the same inconsistency that afflicts the human will, understood as a pure will to will. Human beings remain with their gaze fixed on their lack, which now appears to be at once *both spiritual and natural.* And by this point they have only one task, that of being born, that of bringing their own humanity, *both spiritual and natural,* to completion, a task that is impossible for them to realize, trapped as they are in the infernal circle of their unconscious "longing" for will and life. From now on, human beings will be condemned to "long" for their very humanity, *both spiritual and natural,* without succeeding to reach it. "Hence it is an activity *that has no object whatever,* but is nonetheless *irresistibly driven out* [*getrieben*] *towards one,* and is merely *felt* [*gefühlt*]. Such a determination in the self is called a *longing;* a drive towards something totally unknown, which reveals itself only through a *need,* a *discomfort,* a *void,* which seeks satisfaction, but does not say from whence.—The self feels a longing within itself; it feels itself in want" (265).

2.2.2.1.3. Third step. One needs to understand in what manner this conjunction of nature and spirit, of will and life, under the sign of human defectiveness also connotes the sensible and corporeal aspect of the human figure. If the longing hollows out a void in natural man that distorts his image and consistency, then his corporeal foundation must also bear the traces of the same "discomfort." In other words, the human body will also have to be penetrated and uncompleted by the sharp blade of subjectivity and its negativity. The human body will also have to be shown to be defective, lacking and missing something, like the longing whose natural support it is. And it is precisely for this reason that Fichte makes a subtle but crucial distinction between the "organization" and the "articulation" of the human body. The starting point is still Kant. Fichte says that every living being is an "organized" being, is a whole (*Ganze*) composed of parts, which finds its finality or purposiveness *within itself* (a finality or purposiveness *internal* to the organism). Each "organized natural product," moreover, tends toward its *self*-preservation (*Selbst-erhaltung*), thus tending to conserve that Self, that singular whole-parts configuration, which it embodies. This is the tendency (*Streben*) of natural "bare life." The human body, however, is not only organized like all natural bodies but is also "articulated," Fichte maintains, and articulation is something that differs from organization, it is a kind of intrinsic, congenital disorganization of human organicity: the human body in fact has the ability to change, to articulate its form every time according to the activity undertaken and the purpose the will is aiming to attain. Fichte stresses that this is an indispensable requisite of our freedom, because otherwise we would be bound to completely rigid behavioral patterns, like those to be found in plants or any other animal. In contrast, in us the articulation manages to introduce a margin of indeterminacy that allows the body to shape and modify itself in order to assume different forms, which in turn allow free and diversified activities such as those displayed by humans. Human beings are indeed their own purposes, as are all organized beings, but they are an "infinite purpose," for they are the subject of a pure will, of an absolute will, which cannot be contained in any spiritual or natural form. Therefore the human beings' bodies themselves, not only their spirit, must bear the traces of an irreparable, albeit fortunate, incompleteness. "Articulation does not in turn produce organization, but points instead to another purpose, i.e., articulation is fully comprehended and reduced to a unity only through another concept. This could be the concept of *determinate free movement,* and then the human being would be an *animal.* [. . .] But the human body cannot be understood even through this assumption. Thus the articulation of the human body would have to be such that it could not be comprehended through any *determinate* concept at all. Its articulation would have to point not to some *determinate sphere* of arbitrary movement, as in the case of animals, but rather to all conceivable movements *ad infinitum.* The articulation would not have any determinacy, but only an infinite determinability; it would not be formed in any particular way but would be only formable.—In short all animals are complete and finished; the human being is only intimated and projected. [. . .] Every animal *is* what it is: only the human being is originally nothing at all. He must become what he is to be: and, since he is to be a being for himself, he must become this through himself. Nature completed all of her works; only from the human being did she withdraw her hand, and precisely by doing so, she gave him over to himself. Formability, as such, is the character of humanity. [. . .] To be sure the human being has a plant-like instinct, but he has no animal instinct at all in the meaning given here. [. . .] If the human being is an animal, then he is an utterly incomplete animal, and for that very reason he is not an animal. [. . .] And it is precisely because of this incompleteness that the human being is capable of such formability [*Bildsamkeit*]" (Fichte 2000, 74–78). These are celebrated passages that underlie many later reflections (Gehlen, Plessner, Anders) and mark the birth of modern "anthropology" (79), entirely based on a metaphysics of a defective will/life. What should not be neglected are the consequences that Fichte draws from the human body's "incompleteness" axiom (*Mangel an Vollendung*). These are astonishing yet logical consequences, which cast more light on the profile of modern human beings. If we follow Fichte, we have to concede that corporeal and natural human beings, since they are "longing" and unfinished, do not compose a corporeal and natural whole, an entirety (*Ganze*). That said, every natural organized product should be a whole, for it is in itself as a whole that it finds its internal purposiveness, namely the Self (*Selbst*) of its self-preservation (*Selbst-erhaltung*). And man should be no exception to this rule. Yet, let us now try to turn the argument on its head. Let us ask ourselves: where does a *nonhuman* organized natural product (which is a whole) get the ability to refer to itself? Where does it get that gift of the Self that allows one to speak of its *own* drive to *self*-preservation? In principle, within the framework of Fichte's metaphysics, this subjective nuance of nonhuman nature should also derive from human subjectivity, ergo there should be complete reversibility between the spiritual (human) and the natural (human and nonhuman)—otherwise we would be forced to assume the existence of two types of subjectivity, one of natural life and the other of spiritual life (an assumption that would be incompatible with Fichte's metaphysical premises). So, how can we cover the residual distance between the human and the natural, which ought to coincide point for point? The only way, as strange as it may sound, is that of completely superimposing the idea of human body, which is an *articulated* organic body, to the idea of organic body *in general*. Only in this way, only if the human body is "changed" into the living body in general, into the body of life as such, into the body of nature itself, is it possible to save the principle of *self*-preservation of nonhuman organic bodies, which already contain a seed of subjectivity (human, spiritual). As a result, Fichte concludes: "As has often been pointed out, self-sufficiency, which is our ultimate purpose, consists in everything depending on me, and my not depending on anything, in everything that I will to occur, in my entire sensible world occurring purely and simply because I will for it to occur—just as happens in my body, which is the starting point of my absolute causality. The world must become for me what my body is. This purpose is of course unreachable; but I am nevertheless supposed to draw constantly nearer to it, and thus I am supposed to fashion everything in the sensible world so that it can serve as a means for achieving my final end" (Fichte 2005, 217). *My world has to become my body*. This entails that no natural organized product is, then, really complete, really finished. And this explains why it always remains in the process of *self*-organization, *like a human body, like a spiritual body*. Thus, the *Wechsel* of the category of the autonomous will with the category of the "longing" life tends to break through the fragile walls of the human, *so as to extend itself to bare life, to life as such*. The word *life* begins to acquire a new meaning. We are at the twilight of the ancient teleological vitalism, according to which each living organism obeys a *final form* of its own, preordained for all eternity. We are at the dawn of an altogether different vitalism, on whose horizon life will explode within living beings as a *final force*, blind and savage, that ceaselessly deforms and transforms its profile.

2.2.3. "Life is the autonomy in the phenomenon; it is the scheme of freedom, insofar as it reveals itself in nature" (Schelling 1980b, 222). This is what Schelling writes, in the wake of the results that Fichte was obtaining in those same years. To be precise, this is what we can read in § 9 of the *New Deduction of Natural Right*, which according to many interpreters marks the birth of a Schellingian philosophy no longer nailed to the Fichtean wall of the *Science of Knowledge* (Semerari 1995, XXXII). But how does Schelling arrive at a similar definition? How does he get to the conclusion that freedom and nature ought to be *identified* in the name of life? In this case, it might be useful to directly enter the text and follow its most crucial arguments.

2.2.3.1. The starting point is always given by the Kantian primacy of practical reason over theoretical reason, namely, by the primacy of freedom which is the "unconditional" (*das Unbedingte*) and coincides with the autonomy of one's will (§ 1). Consequently, as Schelling declares at the beginning, "If I am to bring the unconditional into reality, it must cease to be an object for me," in the sense that it is not on the level of theoretical reason and objective knowledge, but rather on the level of practical knowledge and subjective will that I can actually realize the unconditional as something "identical with myself" (§ 2). "Be! In the highest sense of the word; cease to be *yourself* as a phenomenon; endeavor to become a noumenon as such! This is the highest call of all practical philosophy" (§ 3). We are not far away from Kant and Fichte here. But innovations arrive immediately afterward. "If you are a being by yourself, no contrary power can change your status, none can limit your freedom. Therefore, in order to become a being by yourself, *to be absolutely free*, endeavor to subject every heteronomous power to *your* own autonomy, endeavor by freedom to extend your freedom to an absolute, illimitable power" (§ 4). The step is brief but very taut. Schelling contrasts the freedom of the autonomous subject with all heteronomous powers. At the same time he emphasizes that in principle no heteronomous power can limit the autonomy of my freedom, of my will. Hence a strident aporia: this collision between autonomous freedom and heteronomous powers generates contradiction and unleashes a tension at the heart of being; a tension, an "endeavor" (*Streben*) that in turn reveals the true nature of the unconditional, of my autonomous will. The unconditional must be thought of as a kind of call, of demand, or as a pure and simple "commandment" (*Gebot*). "This commandment is unconditional, because it demands something unconditional. Therefore the demanded endeavor itself must be *unconditional,* that is, it must depend only on itself and cannot be determined by any foreign law" (§ 5). This is the reason why, Schelling continues, "By proclaiming myself as a free being, I proclaim myself as a being who determines everything resistant, but is not determined by anything" (§ 6). The key word is "proclaim" (*kündigen*). Autonomous subjectivity is a proclamation, as Schelling himself proclaims; it is a commandment that manifests the peremptory demand of the unconditional. The proclamation of the autonomous will thus becomes the basic axiom around which Schelling's meditations on every other aspect of reality start to revolve. "I rule over the world of objects; even in that world nothing reveals itself but *my* causality. I proclaim myself as master of nature, and I demand that it be absolutely determined by the law of my will. My freedom keeps every object in the bounds of a phenomenon [*Erscheinung*] and thus prescribes to it laws it may not break. Autonomy [*Autonomie*] pertains only to the immutable *Self* [*dem unveränderlichen Selbst*]; everything that is not this *Self*—everything that can become an *object*—is heteronomous, and for me phenomenon. The entire world is my moral property" (§ 7). *Die ganze Welt is mein moralisches Eigentum.* Fichte had already arrived at a similar conclusion, starting from an identical query. How could one recompose the Kantian fracture between phenomenon and noumenon? How could one stitch the realms of noumenal freedom and of phenomenal necessity back together again? Which bridge could allow my autonomous will and an opposite, heteronomous, power that predominates in the natural world to communicate with one another and resolve their mutual tension? The bridge that Schelling decides to cross is one of those that will carry one far. "If I am to rule in the world of phenomena and govern nature in line with moral laws, my causality must reveal itself [*sich offenbart*] through a *physical* causality. Now *freedom* as such can announce itself only through original autonomy. Therefore this physical causality, although it is heteronomous with regard to the *object*, is *autonomous* with regard to its principle, that is, it is not within reach of any natural law. Thus it must unite in itself both autonomy and heteronomy" (§ 8). Each and every word in this paragraph will become a projectile of modernity. *If* the will proclaims its autonomy, *then* this proclamation and this autonomy cannot turn their back on the world of phenomena; on the contrary, they must penetrate into that world, or else they would disappear into thin air. Consequently, hidden somewhere, there must be a channel between the freedom of the autonomous will and the necessity of heteronomous phenomena that blindly, mechanically, follow the deterministic nexuses of cause and effect. This channel is what Schelling defines as a "physical causality," by means of which freedom imposes itself on phenomena and natural laws, a causality that must be both heteronomous and autonomous. But what could this mysterious causality ever be? Schelling's answer follows that given by Kant. "The name of this causality is *life*. Life is the autonomy in the phenomenon; it is the scheme [*Schema*] of freedom, insofar as it reveals itself [*sich offenbart*] in nature. This is why, of necessity, I become a living being" (§ 9). *Ich werde daher nothwendig lebendiges Wesen.* I, who proclaim my autonomy, become a "living being." Each I, each Self, becomes a "living being." Humans become "living beings." Since it is life that now becomes the bearer of my freedom, of the autonomous will that makes all of us truly human.

2.2.3.2. This is an idea that Schelling will never abandon. Here I do not intend to recall the many trajectories of his thought, which in those years roams from medicine to philosophy (Moiso 1990). Let us just mention two more texts. In his *Overview of the Most Recent*

Philosophical Literature (*Allgemeine Übersicht der neuesten philosophischen Literatur*), Schelling sketches out arguments that will then be systematized in the following years. What interests us is the life/will nexus, which is one of the pivots of his budding philosophical system. To begin with, Schelling reiterates that the "spirit," that is, human beings' original autonomy, is the true principle of philosophy. And philosophy for him is a synonym of Kantian philosophy. "It is therefore evident that both Kant's theoretical and practical philosophies are equally incomprehensible and devoid of foundation if they do not derive from a single principle, that of the original autonomy of the human spirit" (Schelling 1988b, 125). *Der Geist will.* The spirit wills. For the spirit is only a will of its-Self, a will that wants itself and eternally flows back into itself, thereby determining itself. "This self-determination of the spirit is called *will.* The spirit *wills,* and is *free.* One cannot provide any further foundation for the fact that it *wills.* In fact, precisely because this action occurs *purely and simply,* it is a *will.* [. . .] This is the action we were searching for since the beginning, the action that unites theoretical and practical philosophy. One cannot provide any further foundation for this action, because the spirit *is* only because it wills, and knows itself only because it *determines itself.* We cannot go beyond this action and this is why it is fully entitled to be the *principle* of our philosophizing. The spirit is an *originary will* [*ein ursprügliches Wollen*]" (121–22). But where can the originary will individuate and determine itself, moving out of its own abstractness? The answer is: in "nature." Only in nature can the indeterminate, noumenal will give itself a determinate, phenomenal content. Accordingly, nature is permeated by the reflexive subjectivity of pure will, whose unstoppable movement of return into itself shapes, molds, configures (*einbilden*) "life" and "living beings," making them alive and kicking. "The spirit must intuit itself as an *object* which has *a principle of movement within itself.* Such a being is called a *living being.* Thus, in nature there is necessarily life. And just as there are a series of degrees in organization, there are a series of degrees in life. The spirit moves closer to itself gradually." In other words, so that the spirit may find itself, so that it may reflect itself and return to itself, expressing its infinite self-determination, it is necessary "that it manifest itself to itself *externally,* and precisely as *organized, animated matter.* Only *life* is, in fact, the visible *analogon* of spiritual being" (115). So, the spirit is "a nature that organizes itself" (113). *Geist* is nature, *Geist* is life. For nature is life—a life that organizes itself and spontaneously tends to perfect its own organization; a life that, consequently, results in a process of dynamic and progressive self-organization of the living being, in which the dynamic and progressive process of self-determination of the originary will expresses itself; a life that, in the final analysis, reveals itself in the will to life, in the will to see the living being grow and flourish, in a peremptory will focused on the organism's health and on its healing from all the traces of inorganicness and dysfunctionality that the vital process, life's *elan* (*Schwung*), on its own, does nothing but contrast. Such is the secret of *life.* "If inside the body there were not a continuous grafting of one function onto the next, a constant reproduction of one on the part of the other, a balance of forces that is continuously reestablished only to be broken again, life would cease" (115). At this point, Schelling declares, we are faced with a new "concept of life." In a footnote, he boasts of his discovery and the fertility of its implications for future research in the medical and biological fields. If up to that moment, he says proudly, there had only been confusion in studies "on the origins and beginnings of animal life," things are now going to change, for "the *concept* of life that I established previously is easily applicable to the *phenomena* of life" (115). As the ensuing history of Romantic biology will abundantly prove, Schelling's enthusiasm was not unfounded. But, apart from this, what are the salient features of this new concept of life, defended by a philosophy that concentrates on "the *living*" (129) and preludes the birth of biology, which takes place precisely in those years? Roughly, we can distinguish three of them: (A) *Purity and priority of life,* in the sense that life is no longer seen in conjunction with its (vital or final) forms; life shimmers as bare life, prior to any form-of-life, passing from the position of defined term to that of defining term (it is life as such that defines living beings, not living beings that define life). (B) *Eternal progressiveness of life,* in the sense that life is now conceived of as a—potentially infinite—process of self-organization of the living creature, which tends to affirm itself more and more and to simultaneously enhance its vitality with the growing self-affirmation of life in its-Self. (C) *Radical defectiveness of life,* in the sense that life, in its unceasing effort at self-organization and self-affirmation, begins to cultivate illness within itself, begins to cultivate the seeds of the inorganic that life, on its own, tends to organize and vitalize. What is the source of this unprecedented concept of life? The concept of pure, originary, absolute will. If on the noumenal level the will proclaims itself to be pure and autonomous, on the phenomenal level life too tends to make itself pure and autonomous, detaching itself from living forms, in order to convert itself into a stand-alone force that produces organisms, instead of deriving from them. What is the result of this unprecedented concept of life? Its perfect "identity" with the concept of will. And this is what Schelling will call, very shortly, "the Absolute."

2.2.3.3. Such is the metaphysical scheme around which the *System of Transcendental Idealism* revolves. There is no need to review the entire series of conceptual equivalences that Schelling lays out starting from the originary will, which is the principle of all his philosophizing. Will is intelligence: this is the point that links theoretical and practical philosophy, and which Schelling here once again uses as his point of departure. Unsurprisingly, and quite consistently with previous texts, he answers his own question as to whether "the intelligence is organic at all": yes, "indeed it is" (Schelling 1978, 122). The continuity of his thought is clear and evident. In the *System* the concept of intelligence (*Intelligenz*) replaces that of spirit (*Geist*). And once the identity of intelligence and nature, will and life, has been established, from this one is meant to conclude that "intelligence is thus an endless endeavor towards self-organization" and that therefore "a graduated sequence of organization will also be necessary" (123). In this regard Schelling speaks of "evolution" (*Evolution*), and he is among the first. "But this scale of organization merely refers to different stages in the evolution of the universe. Precisely as the intelligence, by means of the succession, constantly tries to depict the absolute synthesis, so likewise will organic nature constantly appear as struggling [*strebt*] towards universal organism and at war against an inorganic nature" (125). Said differently, just as the autonomous will tends to eternally return to dwell in itself, each time reaffirming a pure and empty will of the Self, in the potentially infinite series of its Self-representations, so does autonomous life tend to eternally return to dwell in itself, each time reaffirming a pure and empty life of the Self, within the ever more evolved series of its natural organisms. But precisely because life is detached from each form-of-life, precisely because it is conceived of as a vital force that precedes any organic form, life cannot but return to itself *ad infinitum*, continuing on the path of the search of its-Self and making every effort to "evolve" into a universal organism from which every trace of inorganic nature, of a nature opposed to life, will have finally disappeared. This is the reason, for Schelling, *life is essentially sick, defective, lacking*; because life is the expression, never definite and complete, of the Self in nature. "The basic character of life, in particular, will consist in this, that it is a sequence, reverting into itself, fixated, and sustained by an inner principle; and just as intellectual life, whose image it is, or the identity of consciousness, is sustained only by the continuity of presentations, so life is sustained only by the continuity of internal motions; and just as the intelligence, in the succession of its presentations, constantly struggles to achieve consciousness, so life must be thought of as engaged in a constant struggle against the course of nature, or in an endeavor to uphold its identity against the latter" (127). Now, this infinite struggle of life for life, for health, for the affirmation of its-Self, takes place on two levels according to Schelling: that of universal nature (or life) and that of human nature (or life). But the two planes intersect, with human beings at the center of this metaphysical prism—human beings who are doomed to search for their own humanity, wandering between nature and History. What are human beings, in fact? Human beings are without a doubt the most elevated organisms, and those in which nature reaches a peak of organicity and vitality. However, precisely because they are such, precisely because they are nature's organic and vital peak, human beings are also the most indeterminate and inorganic organisms, those in which life manifests itself with maximum force, in its explosive purity, something that wears down any organicity, to push forward and unleash a further, infinite "evolution" of organic form. In this sense nature evolves, nature is in and of itself History, and human beings are those who carry the imprint of this "identity" that is always *in fieri*, in which nature historicizes her vitality. So, if in inferior organisms on the evolutionary scale the force of will/life is still held back by organic forms tied to a purpose, whose behavior is thus predetermined, in human beings the force of the life/will instead explodes into the indeterminate form of finality, devoid of a purpose that is not the purposiveness or the finality in itself, in other words "ideality" as such (215–18). "Nature in its purposive form speaks figuratively to us, says Kant; the interpretation of its cipher yields us the appearance of freedom in ourselves. In the natural product we still find side by side what in free action has been separated for purposes of appearance. Every plant is entirely what it should be; what is free therein is necessary, and what is necessary is free. Man is forever a broken fragment" (216). *Der Mensch ist ein ewiges Bruchstück.* Human beings as a natural kind, human beings as exemplars of the whole of mankind, are "split," "divided" from themselves (*getrennt, entzweitet*). Within the life and the will of human beings, subsumed into the human kind, it is the movement of the Absolute that flows, the movement of continuous return of the Self into itself that constitutes the Absolute. As a result, human beings, in their natural humanity, become "progressive" beings for Schelling, they become a bridge toward themselves, they become a funnel by means of which life, abstracted away from all forms-of-life, gradually realizes its empty identification with the originary will, with the pure will of the Self. In human beings, in the human species, nature reveals itself as History and History as nature, in the name of an infinite "tendency to progress" (*Progressivität*) of the human as such (202). Human beings historicize themselves integrally, because their natural humanity remains a task to be realized. "All my actions, in fact, proceed, as to their final purpose, toward something that can be realized, not by the individual alone, but only *by the entire species*" (205). Hence the tragic destiny from which no one can escape, a destiny foreshadowed by the question that will torment the moderns. The "final purpose" of each and every human being, the purpose that orients the History and the "progress" (*Progressus*) of the "entire genus," what could it possibly be? For Schelling, as is foreseeable, on the one hand it will be absolute autonomy, i.e., *freedom,* on the other hand absolute organicity, i.e., *health*—freedom and health that unfortunately point to the two ideal, by no means real, sides of the Absolute. Thus, given the premises and the development of the entire system of transcendental idealism, on the one hand it is not surprising that "the more an individual acts, the less free does he become" (168), since this is the paradox that underlies the process of inexhaustible self-identification of the Absolute; on the other hand it is not surprising that, by dint of the same paradox, health reveals itself as a chimera for humankind, a purpose without content, a pure purposiveness or finality, a simple "ideality" detached from all reality (129). Human beings, skewered by their membership in the human race, are constitutionally ill. History is nothing but the process of our infinite healing from our natural humanity. We, humans, are nothing but broken fragments. For the human being is "this contradiction [that] strikes at *the ultimate in him*" (222). Therein lies the riddle we are. "Yet the riddle could reveal itself, were we to recognize in it the odyssey of the spirit, which, marvelously deluded, seeks itself, and in seeking, flies from itself" (232).

2.3. *Aseitas.* In modern times, this concept works in an invisible yet decisive manner. In the Middle Ages the term mostly had a negative meaning, that of *sine causa*: God is without cause, is not caused by anything, and in this sense God is *a se ipso.* With Descartes the term begins to lean toward a positive meaning, that of *sui causa*: God is not only without cause, but also cause of Himself. Divine *aseitas* surfaces on the margins of Cartesian subjectivity, which, being finite, cannot found itself but has to be based on infinite divine power, which alone is capable of sustaining the aporia of a self that institutes its-Self. The infinity of a transcendent God allows Descartes to overcome the aporia, otherwise insuperable for a finite mind, of a self that causes its-Self without preexisting its-Self. In order for this to be possible, one has to exit time, the linear sequence of instants that is at the origin of the aporia (how can the Self establish itself without preceding itself?), entering a different dimension, that which corresponds to God's eternal presence—the eternal, inscrutable, omnipotent simultaneity of God with Himself. Finally, with Kant the autonomy of the will becomes the mold, the model, of a new form of *aseitas*, no longer transcendent but immanent. It is subjectivity that now embodies *aseitas*, the aseity of a will that wills itself and thus founds itself, without ever founding itself definitively, but remaining suspended and tied to this process of infinite self-foundation and self-causation. In such a perspective two problems immediately arise. The first is that of an aseity that carves itself out of subjectivity as lack, negativity, defectiveness of the subject or the Self, always in search of its-Self. The second is that of a graft of such an infinite aseity of the autonomous will, that wills itself, onto the sensible and finite world, that represents the other than the Self. The latter problem explains why the autonomy of the will, understood as the aseity of a will *sui causa*, is "reflected" (Kant), "exchanged" (Fichte), and then "identified" (Schelling) with the autonomy of life, understood as the aseity of a life *sui causa*—a life that feeds and perpetuates life for its own sake, detached from all forms-of-life; a life that is abstracted away from every *forma finalis*, thus being reduced to "bare life"; a life that ends up sinking into itself, obscuring itself and falling ill.

2.3.1. With Fichte and Schelling, as seen before, life is permeated by autonomy, transforming itself into a principle of constitution and interpretation, *ratio essendi* and *ratio cognoscendi*, of nature and of human beings. What is changing is not a scientific paradigm. The morphing is metaphysical. In the geography of thought, in the great network of concepts thrown by humans over reality, life now assumes a much more central position. It is not defined by other entities; it rather begins to define other entities that are peripheral to it. It is no longer defined by other coordinates, but it itself becomes a coordinate. Autonomous life and autonomous will become cardinal points of a world no one had previously entered. In the immanence of *aseitas*, life and will come to coincide, to become "identified," thus unleashing an overarching readjustment of the conceptual field that frames the understanding of nature and human beings. For example, one begins to insist on the manifestation of the "living" and the "free" within the "organic." "Organic life becomes the dominant type of free life" (Schlanger 1971, 121), and, almost as if afflicted by an epidemic, philosophy itself becomes an organism (Schlegel) followed by art (Schelling), the law (Savigny), the state (Müller), language (Humboldt), race (Gobineau), the people (Ancillon), civilization (Spengler), and society (Comte, Spencer, Lilienfeld). The category of "organism" starts appearing in all contexts, raising the same paradoxes everywhere. If the organism is the expression of an autonomous and detached life, then the organism will carry the seeds of disease within itself, since autonomous life is *an infinite struggle for life,* a struggle that must continue, that cannot end, on pain of the extinction of life itself. Therefore life is ill, the organism is ill, and both tend to degenerate. "Degeneration imposes itself as something inevitable, and yet one deplores it; it's a natural occurrence, and yet it becomes something one can be accused of" (187). But degeneration is a *natural* process. Why does it become a *moral* accusation? Because here a defect in the autonomous life becomes the echo of a defect in the autonomous will. The two sides of modern autonomy are fully continuous. Within this entirely new framework, the unprecedented *medico-moral* scheme that will underpin the new biopolitics advances, with its vigorous emphases on the morality, the ideality, and the inescapable centrality of natural health that should inspire human action, both in private and public domains. "Hence the importance of medical, hygienical, and pathological issues to be addressed from a naturalist point of view: only against this background one can develop a doctrine of political action. Only against this background can one imagine that it is possible to influence, to bend, to control the course of biopolitical reality without resorting to arbitrary manipulation" (176).

2.3.2. Thus, step by step, health acquires a *normative value.* Health becomes a demand, an imperative, an ideal to which nothing real corresponds. Arthur Schopenhauer is one of the outstanding voices that complain about this unattainable health, which predestines creatures to tremendous illnesses (to be followed not much later by Friedrich Nietzsche). His starting point is the identification of life and will in a world obscurely dominated by our uncontainable yearning for our-Selves. "The will, which, considered purely in itself, is without knowledge, and is merely a blind incessant impulse, as we see it appear in unorganized and vegetable nature and their laws, and also in the vegetative part of our own life, receives through the addition of the world as idea, which is developed in subjection to it, the knowledge of its own willing and what it is that it wills. And this is nothing else than the world as idea, life, precisely as it exists. Therefore we called the phenomenal world the mirror of the will, its objectivity. And since what the will wills is always life, just because life is nothing but the representation of that willing for the idea, it is all one and a mere pleonasm if, instead of simply saying 'the will,' we say 'the will to live.'" (Schopenhauer 1896, 354). For Schopenhauer the world has two sides, the will and life, which are the two sides of a single world, in which a single "will to live" flows. In this perspective the abstraction of life from the living creature is an indisputable fact. One shared life pulsates behind all living beings, and it brings one shared will to the fore. But the problem is now another one, namely the dark consequences of this metaphysical abstraction. If in fact life is conceived as a "will to live," if the natural world is only the mirror of a blind will to life, this means that *life never lives enough*: this is precisely the reason life insists on *willing to live*. And this explains the insistence on pain and illness, on the defect(s) of life, in which the defect(s) of the modern will reverberate. Life suffers, because it has an immense appetite for itself. But what does this ultimately mean? First of all it means that suffering itself, the unhealable affliction of the living, the great appetite of life now becomes the only index of their being-in-life. "This goes so far, that every man who endures a great bodily or mental suffering, indeed everyone who merely performs some physical labor which demands the greatest exertion, in the sweat of his brow and with evident exhaustion, yet with patience and without murmuring, every such man, I say, if we consider him with close attention, appears to us like a sick man who tries a painful cure, and who willingly, and even with satisfaction, endures the suffering it causes him, because he knows that the more he suffers, the more the cause of his disease is affected, and that therefore the present suffering is the measure of his cure" (513). Thus, the more they suffer and the more living beings heal, in the sense that the more they suffer and the more they feel life pulsating within them. This is a trap, however, because "healing," here, instead of leading to health, makes of health itself a demand, an imperative, an ideal that is by definition unattainable. If life as such is reduced to a blind "will to live," there can be no way to satisfy and heal it in this world. Where can one then find peace? Schopenhauer's solution is not that different from that of Nietzsche and many others. The solution is to contrast the great hunger for life with a "great health" (Nietzsche), in which, at this point, one should no longer hear the echo of an impossible natural health but rather the echo of a health that suddenly reacquires its ancient meaning of spiritual salvation. *Salus.* If health cannot be reached on the natural level, since life is reduced to a morbid will to life, the possibility always remains to compensate on the spiritual level, making modernity slam against its own past, creating new religious monsters, making a theological salvation chant reemerge from the biological mutism of natural health. And it is not that important whether it is the Veda (Schopenhauer), Zarathustra (Nietzsche), or still other divinities who announce this salvation. What matters most is that the new gods will inevitably be new demons, whose profile is etched by the idol hidden behind them, the modern idol of Life. This invisible idol will render all those announcements false and derisory. Worse, it will make them ferocious.

2.3.3. Once again it is Schelling, the prophet, who first gives voice to the future. The idea he approaches, some years after the *System,* is that one should be able to turn the ideality of natural health into the reality of spiritual salvation. Said in a slightly different way, the idea is that one can heal modern life, which thirsts for natural wellness, by taking a step backward, and returning to the ancient and venerable tables of religious redemption. The essay in which this operation is carried out, an essay that still garners attention due to its bizarre cross-contamination of the ancient and the modern, of morality and medicine, of old theology and new biology, has a symptomatic title: *Philosophical Investigations into the Essence of Human Freedom*. Freedom is in fact the very symptom of modernity, so long as, with Kant, Fichte, and Schelling, we make it the *fact of autonomy*. And the basic assumption of this essay does nothing but confirm what we already know. *Wollen ist Urseyn*. "Will is primal being" (Schelling 2006, 21). Again, the thesis about the ontological primacy of the will is associated with another one, which represents its corollary in Schelling's view. *Leben ist Urseyn*—life is primal being. "If a philosophy is lacking this living foundation [*lebendiges Fundament*], which is commonly a sign that the ideal principle was originally only weakly at work within it, then it loses itself in those systems whose abstract concepts of aseity [*Aseität*], modifications, etcetera, stand in sharpest contrast with the living force [*Lebenskraft*] and the richness of reality" (26). For Schelling, by contrast, in order to give fullness and concreteness to the abstract concept of aseity one has to immerse oneself in the deep entrails of life, in the "living force," in the force of life—even though, in crossing this threshold, not everything will proceed smoothly.

2.3.3.1. Once the concept of aseity has been injected into the concept of life, the "living force" in fact ends up twisting in on itself, like the originary will. Life becomes an empty will to life, by which life itself—a life that no longer lives enough—is shaped like a disease, a disorder, an infirmity. "Disease which, as the disorder having arisen in nature through the misuse of freedom, is the true counterpart of evil or sin" (34). Moving from the coincidence between will and life, between spirit and nature, we thus run into the problematic coincidence between the old *theological* problem of evil and new *biological* problem of illness. Not that the idea of illness as an abuse and loss of freedom were a true novelty for Schelling, since this was a result he had already arrived at a while beforehand, at least since the years in which he had drafted the *System*. The novelty of his *Investigations* rather consists in following now the same path in reverse, passing from the medical-natural plane of life to the moral-spiritual plane of will, in order to solve the aporias of an insatiable and restless life. This life is the illness of all living beings, humans *in primis,* who for this reason are in urgent need of health. Where can they find it? The solution that Schelling has in mind goes back over the idea of God. Yet, the God he talks about is a new God, different from that of the past. It is a God in which it is necessary to distinguish between the ground (*Grund*) and the existence (*Existenz*). "The natural philosophy of our time has first advanced in science the distinction between being in so far as it exists and being in so far as it is merely the ground of existence" (27). The *Naturphilosophie* that Schelling alludes to is obviously his, and it is a philosophy that for the first time allows one to discern two faces of God, a light one and a dark one, two faces that initially seem to be extraneous to one another.

2.3.3.2. The first face is that of God or of "being insofar as it exists." It is the revealed face of God; it is the face of the God of love and reconciliation, of the person God, incarnate in the figure of Christ. We are here in the presence of the reassuring face of God the creator, who guarantees the overall stability of creation, of the world as an organic and perimetered "system." This God, however, should no longer be seen as the self-evident ground or foundation of Himself, Schelling maintains. Before the God of Christian religion we find something else: we find the dark face of God, which is simultaneously God and before God. At that level we are no longer in the presence "of being in so far as it exists" but rather in the presence "of being in so far as it is merely the ground of existence." Schelling is speaking with no reticence of the very birth of God. Why does God exist? How does God come into being? What is the ground or the reason (*Grund*) for His existence? The answer is given in stages. First of all, Schelling remarks, God exists simply because God *wants* to exist. Will is God's primal being. God wills to exist even before existing. What is more, God exists in the present of His revelation only because in an immemorable past, that follows Him like a shadow, God *keeps on willing* to be. In this sense, the reason for His existence is a "blind will" (32). The reason God exists is a pure desire of Himself, which the revealed God, the existing God, the person God still nourishes in his chest.

Without which, God could not exist. "If we want to bring this way of being closer to us in human terms, we can say: it is the yearning [*Sehnsucht*] the eternal One feels to give birth to itself. The yearning is not the One itself but is after all co-eternal with it. The yearning wants to give birth to God, that is, unfathomable unity, but in this respect there is not yet unity in the yearning itself" (28). The distinction Schelling makes is that between the One and the Self. The One is the person God or the God that exists, whereas the Self is the ground God, the God that wills to exist. Such a divine Self, Schelling says, logically precedes the One—aseity precedes any identity—just as the originary will lurks behind any concrete and personal volition. There could be no better way to cast light on the divide between modern and past ontology. Yet, Schelling avers, all this is not sufficient to explain the existence of God. At this point we must find the bridge that connects God's "yearning" to God's "existence," because otherwise we would not have one single God but two separate divinities. We need to find a metaphysical operator that transforms the originary will into something existent, into a *present and wordly will,* if we don't want one aspect of God to remain detached from the other, thus paving the way to a divine schizophrenia. It is therefore mandatory to change God's drive toward existence into something existing; it is necessary for God's originary will to pass into God's existent will, if we want the ground God to reveal Himself in the person God. But what could a *primal and yet existent will* ever be? In his answer, Schelling does not betray the basic tenets of his past philosophy.

2.3.3.3. The spark of existence in the originary will, the glimmer in which pure will shines and breathes like something existent, is termed a "glimpse of life," that sparkles in God's primordial *will to live.* "Because, namely, this being (of primordial nature) is nothing else than the eternal ground for the existence of God, it must contain within itself, although locked up, the essence of God as a resplendent glimpse of life [*Lebensblick*] in the darkness of the depths" (30). Life peremptorily returns to the center of Schelling's metaphysical apparatus. Life becomes God's primal being, in which His ground and His existence merge. Life is indeed the will that passes into existence, the bridge between the ground God and the person God. Life shimmers as that which *within God* comes *before God.* And it is no longer God that defines life in this perspective, but life that defines God. "God himself is not a system, but rather a life" (62). The metaphysics of life is thus converted into a religion of Life, in which the old and the new, the ancient and the modern, are contaminated under the sign of an unprecedented premonition of what will follow. But, precisely, what will follow? Human beings will follow, to be seen now as the illness of God. If life is thought of as an obscure will to live, that is the ground even for God's existence, this life will always will to live, it will always insist on *willing to live,* without being satisfied with any temporary pause in this or that form-of-life. As a result, the "living force," reduced to a morbid and furious will to live, cannot but empty all living beings of life, because in all living beings life can breathe only as an unsatisfied *will to live.* This is why Schelling speaks of a veil of sadness that is extended over the face of all natural creatures. No living being, from now on, will ever live enough. "Hence the veil of dejection that is spread over all nature, the deep indestructible melancholy of all life" (63). Life is plagued by its-Self, by its *will to live.* All of creation is condemned to the pains of a *will to live* from which an insufficient, defective, and unfulfilled life surfaces. Hence Schelling's desperate query: can God save us from this natural hell? Can we at least hope in God, given that we are no longer allowed to hope in natural health? We could hope in God, in spite of everything, if an ancient and venerable God were still towering in front of us, the God who was formerly worshiped as the enlightened demiurge of creation. But the modern God whose advent Schelling ushers in is a sick and divided God. He is a tormented God. God himself is in fact haunted by a *will to live* that inflames His existence. God himself, Schelling suggests, is in search of His own health. God himself fights for His own healing. For this reason there will be no salvation in God. Not even God is safe and healthy within the horizon of modern life. On the contrary, God is shaken by life, overcome by its imperative, which yells its blind *will to live.* If God is One, insofar as God is person, God is also Self, insofar as God is ground. And between the One and the Self the bridge is life. What does God do then? What does His life consist in? It consists in saving and healing Himself, making His pure, diaphanous, unformed *will to live* shine—a life in which His ground and His existence are rejoined. Such a pure *will to live,* if it continuously saves and heals God from His internal falling out, putting His originary will and His existence in communication with one another in the flame of an amorphous *will to live,* on the other hand condemns all living beings to sacrifice their existence, to be burnt on the altar of a divine *will to live,* without form, without figure. Ergo, humans cannot be saved from the ferocious howl of life, they can only be thrown into its jaws, which devour all that exists. No wonder that at some point Schelling writes that human beings notice the presence of the divine as a "destructive fury" which places them "in an ever higher tension against unity until it arrives at self-destruction and final crisis" (66). For unity is derailed here by the living aseity of the pure *will to live* of the divine. Every transitory identity of life, every form life assumes, cannot but be blown away by the divine wind of life. Because this is how God saves and heals Himself, remaining alive (45–46). "God is life," but this life is now a divine struggle of life for life, for its disruptive abstraction, for its devastating purification from any form-of-life. "God is life," and this means that God himself is struggle because "where there is no struggle there is no life" (63). A struggle from which God will not leave defeated, it will be humans, it will be us, who are forced to immolate ourselves on the purifying pyre (48) of a divinity that infuses life in living beings only as a combustion of their natural form, as an abrasion of their essence and configuration, as illness and torment for their innate insufficiency. In the name of this troubled and belligerent divinity, that is, Life, modern humans will have to raise a battle cry against their own humanity. *Salus vitae.*

3. We have thus arrived at the threshold, at the entrance to the metaphysical world we inhabit, the world of autonomy, in which the horizons of life and will are merged. The two moons that shine over this nocturnal landscape, where the human has been lost, are Health and Salvation. On its starry vault the "great, eternal, iron laws" of life (Haeckel 1934, 310) are written. On a plateau, a sick man contemplates the spectacle below. His name is Charles Darwin. He is a naturalist who, like many of his predecessors, has a passion for "metaphysical speculations."

3.1. What is life? By retracing the answers given to this question before Darwin, who is undoubtedly the most influential philosopher, and not only scientist, of the last two centuries, one is struck by the fact that life has almost always been conceived of on the model of the will. When vital phenomena have been isolated as independent phenomena, distinguishing the organic from the inorganic, the anthropomorphic imprint of the will has almost always been stamped on them. Thus, the notion of vitality has been associated with the notion of finality, with the idea of a purpose, *which by definition is the object of a will.* As already said, the purpose is indeed the recipient of the will; it is that which gives the will a certain form and orientation. But the purpose can also be seen and has often been thought of as the recipient of life, as that which gives life a certain form and orientation. In this case, the purpose turns into a synonym of form-of-life, to be understood as both *forma finalis* and *forma vitalis* (Gilson 1984; Pichot 1993). In our tradition, this is the way vital phenomena have usually been interpreted.

3.1.1. The ancients already conceived of life in this fashion, in the tracks of Aristotle and Galen. For them life is imbued with finality, life is the sign of a purpose, the purpose according to which the living being lives and is shaped, planned, designed. In this regard, it doesn't matter whether the purpose is immanent or transcendent, whether it is imprinted in the matter of living beings (Aristotle) or imprinted in the mind of an otherwordly God (Galen). It doesn't matter because in any case life follows a purpose, it is endowed with will, whatever the theoretical perspective we may adopt. It doesn't matter because we can always associate a living being with a specific purpose, be it immanent or transcendent.

3.1.2. It is true, however, that this view of life, which remains the prevailing one up to the dawn of modern times and resurfaces in the works of some contemporary biologists (Driesch and his school, for example), was not the only one to be put forward in ancient times. Lucretius, following Epicurus, is among the first to denounce the illusion and to contend that life is not imbued with finality but rather appended to chance. Just like the other natural marvels, living beings are produced, he says, by the blind and casual movement of atoms that are not oriented in any way. Yet, the problem is that any threshold between vital and nonvital phenomena tends to disappear from this perspective.

3.1.3. This is also the threshold that tends to be removed by Cartesian philosophy. Descartes's mechanicist conjecture aims precisely to eliminate all purposes, all traces of finality from the natural realm. But the problem, again, is that, once finality has been abolished, all borders between vital and mechanical phenomena tend to vanish, just as in Epicuraneism. The animal-machine is born. The living organism is reduced to an automaton. And human beings, who are the only animals worthy of interest, now seem to be divided in two. On the one side, there is their extended body, which can be described in mechanistic terms. On the other, there is their thought, which is a synonym of the will. Having eradicated all purposes from the realm of the *res extensa,* the framework for vital phenomena thus becomes for Descartes the framework of a still life, from which one can exit only by passing through a mysterious door, the door of the pineal gland, which opens up the paradise of the *res cogitans* to human beings.

3.1.4. With the growth and refinement of naturalistic observations—in the years that follows the birth of the "new science," emancipating itself ever more from Scholastic theology (and teleology)—one notices the insufficiencies of the Cartesian conjecture, which is incapable of *saving the phenomena,* that is, the distinction between organic and inorganic phenomena. Thus, in order not to fall back on the finality-vitality equation, another conjecture is put forth. Newton has recently introduced the idea of a universal attraction between physical bodies, supported by his famous *hypotheses non fingo.* In brief, according to Newton, no one can actually say what the *nature* of such a universal attraction is, and yet this attraction between bodies exists, because we can decode its regularity. In the light of this, there is no need to grasp what gravitational force really is. We can confine ourselves *to looking into its effects.* A task that is more than sufficient for scientific investigation, not only in physics but also in biology. This at least is the idea defended by John T. Needham (a long-time collaborator of Buffon) in the eighteenth century. Needham in fact imagines that vital phenomena can be distinguished from all others due to the presence of a "vegetative force," analogous in every respect to Newton's gravitational attraction. In his opinion, this force is physical, not supernatural. And yet it is not reducible to the forces postulated by Cartesian mechanism. It is rather an "expansive force" that we can attribute to organic particles so as to explain their build-up and integration into ever more complex organisms—an expansive force that we can therefore contrast with the inertia of a "resisting force," inherent in inorganic matter. Vegetative expansion against material inertia. Organic force against inorganic resistance. That is all Needham has to say, almost as if the formulation of the problem could coincide with its resolution. He seems to be content with the fact that vital phenomena can be told apart from all others, since a vegetative (or expansive) force is at work in them whose regular effects can be described. But how can one better specify the characteristics of such a force, which should be seen as an exclusive property of organic matter? Why does one portion of matter possess this force while another does not? What do we mean by vitality? At the time, the question had started to tickle everyone's imagination. That said, it was difficult to move beyond the point Needham had reached. Caspar-Friedrich Wolff for instance, issuing his *Theoria generationis* in 1759, sought to recast the idea of a vegetative force *à la* Needham, to be now understood as a *vis essentialis*. Johann F. Blumenbach, a naturalist quoted by Kant in the *Critique of Judgment*, did the same thing in 1781, transforming Needham's vegetative force into a mysterious *nisus formativus* (*Bildungstrieb*), a drive toward formation and organization that ought to pervade all organic matter. Unfailingly, the model of all these reformulations of the problem of life (which did not provide a solution to it) remained the Newtonian theory of the universal attraction between physical bodies, dressed in his categorical refusal to enunciate metaphysical hypotheses. What is the *nature* of vital phenomena and forces? "It doesn't matter," was the usual answer in those days. Let us just analyze the regular effects of those forces, because we cannot proceed beyond this, and after all we do not need to proceed beyond this. But then, one could object, what are we talking about? What is the *subject* of our investigation? Metaphysical hypotheses, chased out through the door, will sooner or later come back in through the window.

3.1.5. Needham's theses should also be read, and evaluated, in light of the rather fierce debate that saw two alternative hypotheses about the organism's development—at that time termed "evolution"—confronting one another. On the one hand, there was the preformationist hypothesis, according to which the organism's seed or the embryo already contains all its parts within itself (and the seeds of the current generation can already "contain" the seeds of generations to come: the *emboîtement* thesis). On the other hand, there was the epigenetic thesis, according to which the organism, during growth, does not merely unfold its various parts, whose dimensions should increase with the passing of time, but develops, or "evolves," in a different manner (hence the seeds of today do not "contain" the seeds of tomorrow: a critique of *emboîtement*). It is in the course of this debate that the idea of a "spontaneous generation" of living beings emerges. The idea, in brief, is that the living being, the organism, can emerge from what at first seems to be an inert and disorganized matter: it is on this issue that several of Needham's empirical observations focused. If we believe that processes of gradual and regular aggregation of very minute organic particles, dissolved and dispersed among dead and inorganic matter, are hidden within vital processes, we can think both of a spontaneous generation and of an epigenesis of living beings. In other words, we can think of a sort of slow "construction" of living beings, which is not finalistically prefigured and predetermined by the form of the embryos. We are thus suddenly faced with the problem of a *vitality* that is detached from every *form-of-life*. In fact, so long as we move within a preformationist horizon, life remains subjected to final forms, to vital forms, which are as if stamped and imprinted for eternity in the natural realm. Life simply amounts to living beings, to those vital/final forms in which life as such is resolved and broken up. And there is no room for a life that cannot be reduced to its forms, that cannot be subdivided into them. On the other hand, so long as we move within a mechanistic horizon, the problem of a vitality detached from all forms-of-life does not even arise, since any border between the organic and the inorganic dissolves. Vital phenomena are sucked back into the realm of mechanical phenomena, living beings are reduced to automata. In this perspective, too, there is no room for life as such. But let us now assume that we don't want to accept either option. Refusing both the subdivision of life into its various forms and its eclipse within mechanical phenomena, we will need to find a definition of life, or of vitality, that can preserve its specificity in regards to other spheres of the material world, without appealing to final forms, to vital forms, which *preform* life itself. Hence the queries, the hesitations, the Pindaric flights of those, like the believers in epigenesis and/or in spontaneous generation (Needham, Wolff, Blumenbach, and many more), who imagine that vitality is irreducible to preestablished forms-of-life, but is rather an abstract force independent of organic structures, a force that follows laws of its own to which those organic structures owe their existence and their very configuration. A road one can take, to avoid embarrassment and cut the Gordian knot, is that of a weak animism, which returns to the ancient definition of life as *psyche*, as a vital *pneuma* that breathes in the living. This is the road that some of the major naturalists of the first half of the eighteenth century took, absent other viable alternatives.

3.1.6. This road had been opened two-thousand years previously by Plato, and it will be the one taken by all later platonists or neoplatonists—including those twentieth-century biologists who at all costs wanted to reintroduce the "subject" into biology (Goldstein, Weizsäcker, Uexküll . . .). In this case, the idea is that living matter is animated, namely, that it contains some properties that we usually reserve to the psyche, i.e., to human subjectivity. This is the idea defended by Pierre-Louis Moreau de Maupertuis, for example, who thinks we should preserve the distinction between vital and mechanical phenomena but also distance ourselves from the preformationist thesis (that of Malebranche, Leibniz, Bonnet, Boerhave, Hoffmann, Haller, according to whom life can be subdivided into final/vital forms that fully explain the life of all living beings). Maupertuis instead maintains that organic matter is composed of minuscule particles, which tend to aggregate in more complex organisms by virtue of an intrinsic *vitality,* and which are endowed from the beginning with qualities such as *desire, aversion,* and *memory.* We are far away from Cartesian mechanism here, just as we are far away from the hypothesis of a God who stamps matter with the various final/vital forms of the living. Rather, Maupertuis tends to associate vital phenomena with psychic phenomena. "A uniform and blind attraction, diffused in all parts of matter, is not sufficient to explain how these parts manage to compose a body, even one endowed with the simplest organization. If all parts have the same force, the same tendency to unite, why do some of them go to build an eye while others go to build an ear? Why such a marvelous accommodation? Why do they not assemble in a disorganized fashion? If we want to say something sensible in this regard by means of analogy, we should say that in them there exists some principle of intelligence, or something analogous to what we call *desire, aversion, memory*" (Maupertuis 2005, 205). In this perspective, it would seem that organic bodies are composed of minute particles, of minute animals that are already animated and precisely for this reason, precisely because they desire one another, avoid one another, perceive one another, and remember one another, proceed to compose more complex animals, which in turn desire, avoid, perceive, and remember one another (219). As one can easily see, by embracing such a conjecture, we immediately cross over into animism, into hylozoism, into panpsychism, which at the time was a creed shared by top scientists, like the chemist and physiologist Georg-Ernst Stahl. A creed that inevitably runs into problems: *(A)* In the first place, from this point of view it becomes almost impossible to draw a clear line between vital and nonvital phenomena, because life starts to gush forth everywhere. One cannot *a priori* exclude its presence in any particle of matter, even that which at first sight appears to be inert and inorganic. The entire material world turns into an immense proteiform animal. *(B)* In the second place, we do certainly free ourselves from finalistic preformationism without precipitating into Cartesian mechanism, but only at the cost of endorsing an equally embarrassing thesis, animism, which attributes a hidden will to the smallest elements of the natural world in the form of memory, desire, and aversion. Consequently we do not exit from the anthropomorphic projection of the will onto vital phenomena. We do not escape the metaphysical anamorphosis that transforms life into will and vice versa. "Undeniably, the intelligence we feel within us points to a source from which the intelligence of human beings, of animals, and that of all other beings emanates, each one to the degree appropriate to it" (221).

3.1.7. Although fascinated by Maupertuis's idea, Denis Diderot considers it excessive. As he quickly objects, if we push this idea to its extreme consequences, we reach the conclusion that "the world has a soul as though it were some great animal" (Diderot 1999, 67). Plato waits around the corner. And Plato is too much. That said, the theory proposed by Georges-Louis Leclerc de Buffon, the best known French naturalist of the period, does not represent a more convincing solution in the eyes of Diderot. In fact, Buffon speaks, against the preformationists, of epigenesis and spontaneous generation, but then he ends up speculating about mysterious *moules intérieurs* (internal matrixes) that supposedly govern the organism's development and, in addition to this, ensure the fixity of natural species. As Diderot keenly remarks, by going down this road we fall back into a sort of Aristotelianism, and it seems difficult to combine this renewed Aristotelianism with the hypothesis of a composition of organic bodies that starts with small living molecules—a hypothesis endorsed by Buffon himself in order to save the distinction between living and dead, inert, matter. "Are forms produced according to a matrix? What is a matrix? Is it a real, pre-existing entity? [. . .] If it is indeed a real, pre-existing entity, how was it formed?" (77). The solution to the problem of vitality that Diderot, for his part, provides is that of a slight yet beneficial weakening of the conjecture that had been put forth by Maupertuis. Nevertheless, in reading his words, one can't help sensing his deep embarrassment. Let us no longer talk, Diderot says, about memory and intelligence at the lowest levels of the Great Chain of Being, let us only talk about a diffuse "sensitivity" in all of matter, even in its smallest elements. We can easily be content "to suppose that they [i.e., the smallest organic molecules] were capable only of feelings a thousand times less intense than those which the Almighty has bestowed upon the stupidest creatures who are closest to lifeless matter. As a result of this subdued sensitivity and the difference in configuration, there would never have been more than one position to suit any given organic molecule, and that position would have been the most comfortable of all; this is the position it would constantly have sought, with an unthinking restlessness, just as when animals stir in their sleep, and the use of almost all their faculties is suspended, until they find the position most conducive to their rest" (68). But what is this "subdued sensitivity"? What is this "unthinking restlessness" (*inquiétude automate*) that is meant to qualify and define the life of living beings? Diderot doesn't specify, leaving to his interpreters the task of coming to terms with this paradoxical synthesis of restlessness and unthinking automatism. What is certain is that Diderot's *inquiétude* closely resembles the *uneasiness* Locke had already focused on (and the *Unruhe* Leibniz had referred to immediately after Locke)—a "restlessness" that, according to Locke, is nothing but an attribute of the human will, "restless" at the moment in which it readies itself to decide and eventuates in a concrete act of volition. In brief, not even with Diderot do we escape the anthropomorphic analogy between vitality and the human will. As a matter fact, the *inquiétude automate* he speaks of seems to be only and exclusively that of the naturalist confronted by the enigma of life.

3.1.8. In the very midst of this fiery discussion on the secret of life, one of the most famous and suggestive definitions, the one offered by Xavier Bichat, all of a sudden resounds. "Life consists in the sum of functions by which death is resisted" (Bichat 1978, 10). Rivers of ink have been poured over this formula. But does it really contain a definition of life? Or does it do nothing more than rephrase the enigma? If we stay close to the texts we must certainly opt for the second alternative, since Bichat confines himself to defining life only in the negative: *life is nondeath*. And the most impressive pages of his *Physiological Researches on Life and Death* are in effect devoted to death, rather than life. Bichat does indeed talk about a "vital principle," as had others before him (Barthez, Bordeu), but without defining its nature more precisely, just like those who preceded him. He in fact believes it is not his task to undertake. In his case the epistemological model still remains that of Newtonian gravitation and its correlated *hypotheses non fingo*. We can therefore distinguish, as Bichat does, organic from inorganic matter, assigning a precise list of properties to the former that the latter does not possess, such as variability and unforeseeability—which will prevent us from applying the same mathematical formulae that we use to decipher physical and chemical phenomena to vital phenomena. We can also distinguish "animal life" and "organic life," attributing sensibility and motility to the former, and nourishment (composition) and excretion (decomposition) to the latter. We can then distinguish an animal sensibility, which gathers and centralizes the brain's stimuli, from an organic sensibility, in which the stimuli remain localized at the periphery. And from this we can even conclude that animal life is characterized by simultaneity (sensibility increases with increases in motility) and symmetry (two eyes, two ears, two arms, two legs), whereas organic (or vegetative) life is characterized by alternation (processes of composition alternate with those of decomposition) and by asymmetry (a single heart, a single liver). We can do this and more. Having said that, the fact remains that for Bichat the "principle" of the phenomena under examination is a mysterious and inscrutable agent, and we cannot lift the veil covering it. As a result, we must limit ourselves to examining the manifestations of this principle, the life principle, without asking ourselves about its nature. And it is for this reason that life can only be defined as the opposite of death. "This principle is that of life; unknown in its nature, it can only be appreciated by its phenomena [. . .] The greater number of Physicians who have written upon the vital properties, have begun by researches on their principle, have endeavored to descend from the knowledge of the nature of this principle to that of its phenomena, instead of ascending from observation to theory. The Archaeus of Van Helmont, the soul of Stahl, the vital principle of Barthez, the vital power of others, have each in their turn been considered as the sole center of every action possessing the character of vitality, have each in their turn been made the common base of every physiological explanation. But these bases have every one of them been sapped, and in the midst of their wrecks have remained the facts alone which rigorous experiment has furnished upon the subject of sensibility and motility. So narrow indeed are the limits of the human understanding, that the knowledge of first causes has almost always been interdicted. The veil that covers them envelops with its innumerable folds whoever attempts to rend it. [. . .] In the study of animals let us proceed as modern metaphysicians have done in that of understanding. Let us suppose causes, and attach ourselves to their general results" (10, 76–77). One could argue that Bichat read some philosophy. But he certainly hadn't grasped all the potential and explosive consequences of "modern metaphysics." This explains the ironic and disdainful judgment of him by someone who, instead, had been educated in Schelling's severe philosophical school. "The physiologist has luminously explained Y plus X by informing us that it is a somewhat that is the antithesis of Y minus X; and if we ask, what then is Y-X? the answer is, the antithesis of Y+X. [. . .] The definitions themselves will best illustrate our meaning. I will begin with that given by Bichat. 'Life is the sum of all the functions by which death is resisted,' in which I have in vain endeavored to discover any other meaning than that life consists in being able to live" (Coleridge 1848, 21–22).

3.2. What is life? The problem becomes pressing once things don't add up. This is the moment of modernity, which is not so much a historical moment as a metaphysical moment. It is the moment in which the theological and pyramidal order of the cosmos, where everything was held together by the mind and will of a God, a principle, an ultimate foundation, collapses. Up to that moment one could still craft one's life according to divine will, making living beings into so many purposes of an otherwordly mind. Or perhaps—but not much changes—one could glimpse the pantheistic expression of a World Soul in living beings, for that Soul is also capable of expressing purposes. In any case the problem of a standalone life, of an abstract life-force independent of individual living beings or of One Living Being, had not been posed. The phenomenon of life was coextensive with the dictates of a superior will (God or World Soul), which wisely moved the cosmos. Thus, life was given a precise form and a precise orientation, in which one could still discern the pattern of an intelligent design. The problem, in this perspective, was only that of climbing step after step up the "Great Chain of Being" (Lovejoy 1936), according to which the life of nature was sequenced. Human beings just had to conform to it. This was the metaphysical world before Kant, a world of heteronomous lives and wills, a world where both life and will were subordinated to something higher, a world nailed to eternal forms (or essences), to the unalterable Form (or the Essence) of the One.

3.2.1. The idea of the Great Chain of Being is the idea of a closed cosmos, with a high and a low, between whose extremes all creatures are aligned step by step, continuously. The principles behind a cosmos made this way are roughly two: the *principle of gradualism,* thanks to which one gradually rises from the most imperfect being to the supreme and most perfect being; the *principle of plenitude,* thanks to which there are no intervals between one degree of being and the next. Nature's forms are thus distributed on an ontological scale, which progresses from the low to the high, the *scala naturae,* a ladder that knows no voids, because nature all together is a whole, it is the Form of forms, it is a compact universe, afflicted by *horror vacui.* For centuries thought never left this universe. Let us take Pico della Mirandola, for example, who is sometimes seen as a precursor of the hypothesis (in truth a much more recent one) of a congenital defectiveness of human beings. Pico's world fully satisfies the requisites of the Great Chain of Being. It is a closed, static, and perfect world, embraced and perimetered by an omnipotent mind. There is no standalone life in Pico's *Oration* [*Discorso*], only forms-of-life, arranged from low to high. And his question is the following: what is the *dignitas* of human beings? What is the "place" of human beings on the ladder of nature? The "dignity" Pico speaks of is, in fact, simply the "place," the rank (*dignitas*) of creatures in a cosmos that is hierarchically ordered. Therefore, Pico asks himself, what is the rank of human beings? If human beings had no rank, there would be a hole in the *scala naturae,* and this would open an intolerable crack in the cosmos. And yet God, at the end of creation, seems embarrassed. It seems, Pico tells us, that He is not able to find a "place," a rank, a *dignitas* for human beings. It seems He may have exhausted the "archetypes" at his disposal and is not able to give a "form," a "visage," to his latest creature. "All space was already filled; all things had been distributed in the highest, the middle, and the lowest orders" (Pico della Mirandola 1956, 6, § 4). God, Pico continues, then decides to address the first human, this bizarre "creature of indeterminate image" (*indiscretae opus imaginis*), in order to propose a pact to him: "We have given you, Oh Adam, no visage proper to yourself [*propriam faciem*], nor any endowment properly your own [*munus ullum peculiare*], in order that whatever place, whatever form, whatever gifts you may, with premeditation, select, these same you may have and possess through your own judgment and decision. The nature of all other creatures is defined and restricted within laws which We have laid down; you, by contrast, impeded by no such restrictions, may, by your own free will, to whose custody We have assigned you, trace for yourself the lineaments of your own nature. I have placed you at the very center of the world, so that from that vantage point you may with greater ease glance round about you on all that the world contains. We have made you a creature neither of heaven nor of earth, neither mortal nor immortal, in order that you may, as the free and proud shaper of your own being [*plastes et fictor*], fashion yourself in the form [*formam*] you may prefer" (7, § 5). What does Pico intend to convey in this celebrated passage from his *Oration*? Does he perhaps mean to tell us that human beings are sovereignly free to hunt for their own humanity, to chase it, in perfect autonomy? If this were the case, Pico would without a doubt be a close ancestor of Kant; he would be the first interpreter of the autonomous will, namely, the first to imagine a human will that is completely detached from God and the natural world; he would be the first to conceive of the human will as a will of its-Self that excavates its own void even within its own phenomenal support, the human body, reducing it to an organism lacking an *imago*. The "creature of indeterminate image" to which Pico alludes would already coincide with the human beings that Fichte and Schelling have in mind: natural creatures who do not dispose of a form, of a preestablished organization, but must search for it themselves, without finding it ready made in God or the natural world. And Pico would actually be a prophet of modern philosophical anthropology. But is it legitimate to attribute a similar vision to him? In reality this would seem rash. As a matter of fact, his *Oration* belongs to a different metaphysical horizon. Pico's world is a premodern world, still enclosed within the immense "harmony of the universe," in the Great Chain of Being, from which human beings are not sufficiently autonomous to emancipate themselves. And this is why, immediately after telling him that he will not have "a visage proper to yourself," Pico's God explains to Adam, the first man: "You will be able to degenerate into inferior beings, the brutes; you will be able to regenerate, following your decision, into superior beings, which are divine." What Pico is describing here, with a few masterful strokes, is a spiritual itinerary: human beings have no predetermined "place" in the *scala naturae*, in the hierarchy of natural forms, but this does not mean man can find his "place," his rank, his humanity, elsewhere. Human beings don't have a predetermined "place" only because they can occupy any "place," from the lowest to the highest, only because they can take on any one of the "forms" disseminated in the natural world by the mind of the Almighty. Human beings, however, cannot decide, and not even ask themselves, in full autonomy, about their own humanity; they can only opt for one of the "forms" that God, from the beginning, has imposed on nature. "But upon man, at the moment of his creation, God bestowed seeds pregnant with all possibilities, the germs of every form of life [*omnigenae vitae germina*]. Whichever of these a man shall cultivate, the same will mature and bear fruit in him. If vegetative, he will become a plant; if sensual, he will become brutish; if rational, he will reveal himself a heavenly being; if intellectual, he will be an angel and the son of God. And if, dissatisfied with the lot of all creatures, he should recollect himself in the center of his own unity, he will there, become one spirit with God, in the solitary darkness of the Father, Who is set above all things, himself transcend all creatures" (8–9, § 6). In these adoring words there is no trace of an autonomous will. The will of human beings can only obey God's will, to such an extent as to coincide with it, on top of the ontological pyramid. As regards natural life, it is clearly subdivided into the "forms of life," that human beings can by no means move beyond. Human beings can only choose which of the forms-of-life they prefer, whether a high one or a low one (Bori 2000). Ergo, human beings are not champions of autonomy. For Pico, their deeper Humanity transpires instead in the visage of an otherworldly Divinity. This view of the human being certainly did not represent a radical novelty in those days, and what is more, it will continue to echo for a very long time. Three centuries later, as we have seen before, Herder was still rephrasing it.

3.2.2. Along the Pico–Herder axis an ancient idea of humanity is thus prolonged, one that tends to make of man the "prototype" (Herder's definition) of all life forms. The metaphysical world is still a closed cosmos, entirely perimetered by God's mind. In His transcendent visage, human beings, if they elevate themselves sufficiently, can even come to mirror themselves. But what happens when they do so? What happens when human beings, fully adhering to their own humanity, reflect themselves in the Creator? Briefly said, it happens that human beings reveal themselves as the "prototype" of the entire creation, on a par with the divinity. For not only in God the creator but also in human beings—as Pico and Herder remind us in unison—the germs and seeds of all nature are deposited. Human beings who are really human, that is, really divine, eventually appear to be the "prototype" of all living creatures, the Form of all forms, the *telos* of all natural essences. In human beings, creatures in which it seems as if all others were sketched, what we may call the *teleological prototype* of nature comes to light. The final form of every natural form.

3.2.3. This is not the only variant of the "natural prototype." There exist others, for instance that of the *etiological prototype*. In this case the "prototype" is at the base instead of at the peak, of the natural pyramid. The first to talk about it is Buffon who introduces the notion of "prototype" when confronting the problem of the origin and fixation of natural species. How, exactly, were natural species formed? Via spontaneous generation, Buffon answers, starting from "living matter," i.e., from organic molecules that were initially dispersed in inert matter. At a certain point those molecules were mixed and combined, thereby composing the *first* individual of each natural species, of each organic type, of each form of life. This is what Buffon means by "prototype": the *first* specimen of every species, outlining its "internal matrix" (*moule intérieur*), that which will then be inherited by the following generations of the same species. In this way, Buffon could successfully explain the fixity of natural species without resorting to the mysterious theory of *emboîtement*. Yet, as already seen, he did remain tied, hands and feet, to an Aristotelian metaphysical scheme, according to which life from its inception takes a specific form, and never abandons it. For Buffon, in this sense, life is not yet autonomous and standalone; it is not yet a force detached from its many and diverse forms; it is not yet something distinct and separate from living beings. Those who seemingly get closer to such a conjecture are rather Maupertuis and Diderot. For them the "prototype" no longer refers to the original specimen of *a single* species but to the original specimen of *all* natural species. Both in fact imagine that natural species have a history, that natural species have replaced one another, and that this history began with one universal "prototype." Given this premise, both can thus talk about a hypothetical "transformation" of species in the course of time, and both can speak of a hypothetical "common ancestry" of species, all stemming from a very old, solitary ancestor. It is to Maupertuis and Diderot, therefore, that we must ascribe paternity of these two crucial hypotheses: transformation of species and common ancestry.

"If we consider the animal kingdom, and observe that, among the quadrupeds, every single one possesses functions and bodily parts—especially internal organs—fully resembling those of any other quadruped, is it not easy to believe that in the beginning there was only a single animal which served as a prototype for all the others, and that all nature had done is to lengthen, shorten, alter, multiply, or eliminate certain organs?" (Diderot 1999, 40).

3.2.4. In this perspective, however, the question immediately arose as to how one form-of-life is "transformed" into the following one. This was the question of *vitality*, to be understood as an *active cause* of the transformation of forms, of species. A question that, in point of fact, neither Diderot nor Maupertuis nor others in those days were capable of confronting. If at the beginning of natural history and at the basis of the Great Chain of Being there was a single prototypical Form, by virtue of what had this Form been transformed into new ones? A glimmer of an answer was given by imagining a recombination of the organic molecules in the process of "generation," a conjecture that was put forth by both Diderot and Maupertuis. But this didn't explain a lot ultimately, since one then needed to specify what the laws of this recombination were, and to do this, one needed again to resort to the idea of a life-force detached from the forms-of-life, which ended up taking the shape of an enigmatic vegetative force (Needham), of an enigmatic psychic force (Maupertuis), or of an enigmatic "unthinking restlessness" (Diderot). This is probably the reason, at the time, life itself (or vitality) was thought of as an original Form, as the urform of all possible life-forms. The reason is that it was difficult to conceptualize a vital force abstracted away from all vital forms. It was much more feasible to describe life itself as a proto-form of which all other forms of life were so many "metamorphoses." This is the path Diderot actually took, a path that led him, almost inadvertently, from the notion of etiological prototype to the notion of *morphological prototype*. And this is the path that Johann Wolfgang Goethe also took a few years later. In 1790, in fact, Goethe publishes the great manifesto of naturalistic morphology, *The Metamorphosis of Plants,* where he fantasizes that all plants be the metamorphoses of one single plant, of a primordial plant-form, identified with the leaf-form. *Alles ist Blatt,* "everything is leaf" in the garden of vegetable life. Within a short period of time Goethe extends this hypothesis to the animal world, where "everything is vertebra." The prototypical figure, in both cases, leaf or vertebra, that which Goethe calls the "primal phenomenon" (*Urphänomen*), changes ceaselessly, obeying the two basic laws of "polarity" (*Polarität*) and of "ascension" (*Steigerung*) that govern the cycle of morphogenesis: the cycle of eternal remodulation of the primal Form (Hadot 2004, 225–29, 251–64). In this way, Goethe hoped to confer a less opaque meaning to the concept (previously devised by Blumenbach) of *Bildungstrieb, nisus formativus* or formative force. Unfortunately, with the theoretical tools of naturalistic morphology he could not go that far. For he was not in a position to clearly distinguish between the laws of life itself and the laws of living form(s). So, even though his theses were destined to make Romantic biology blossom, on a level with those of his friend and contemporary Schelling, Goethe's metaphysics seemed to remain one step behind, tied as it was to a reign of natural forms held together by the *vis centripeta* of a proto-form. In the final analysis, although he could smell it in the air and could see its first bursts of light explode in Schelling, Goethe was scared by the idea of an autonomous life; he was scared by a life that has no shape and can pulsate in the absence of any form. This was precisely the "danger" that morphology had to avert at all costs. "The idea of metamorphosis deserves great reverence, but it is also a most dangerous gift from above. It leads to formlessness, it destroys knowledge, dissolves it. It is like a *vis centrifuga,* and would be lost in the infinite if it had no counterweight; here I mean the drive for specific character, the stubborn persistence of things which have finally attained reality. This is a *vis centripeta* which remains basically untouched by any external factor" (Goethe 1988, 43).

3.2.5. To sum up, in the second half of the eighteenth century there are three variants of the "prototype" that make it possible to contain life within the confines of Form: the teleological, the etiological, and the morphological. The overall metaphysical frame of naturalistic investigation still remains that of the Great Chain of Being. And the researches of Jean-Baptiste Lamarck are to be included within this frame. The inventor of the word "biology," a collaborator (as was Needham) of Buffon, Lamarck firmly believes that vital processes should be kept apart from other natural phenomena, but he does not believe that a mysterious "vital force" is at work in living beings. He is rather convinced that living beings are characterized by organization: living beings are bodies articulated into parts. Organization or articulation (the two terms being synonymous for Lamarck) of the living being is the sole index of its vitality. Thus, depending on the greater or lesser degree of articulation of the organic body, we can talk of more or less vital organisms and also of more or less "perfect" living beings. Living beings are indeed articulated forms, whose perfection increases with the increase in articulation. The more living beings are organized, the more perfect they are. Accordingly, life can be broken down into organic forms, which in turn can be arranged along a scale of increasing perfection; that is the scale of increasing organization of living beings. Hence Lamarck's famous idea: the ascending steps of the *scala naturae* correspond to so many evolutionary steps. As everyone knows, in fact, while sticking to the principle of the *scala naturae,* Lamarck does not assume that natural species are eternal forms or essences. He is a nominalist: there undoubtedly exist classes of similar individuals, that is, "species," but "species" can change their properties over the course of time. Over the course of generations organic forms mutate, and each generation of living beings leaves the new characteristics they have acquired in life to the following one. The hypothesis of the heritability of acquired characteristics was quite current at the time: Buffon had embraced it and both Maupertuis and Diderot had made use of it to stutter the first timid hints at the hypothesis of a transformation of the species. With Lamarck, the two ideas, the evolution of species and the heritability of acquired characteristics, are placed within a narrative with quite a different level of consistency. He says that at the base of the natural ladder we find the most elementary and imperfect organisms, infusoria, which arise by means of "spontaneous generation" from inorganic matter (Needham *docet*). Those bodies however don't stay still; they tend on the contrary to become ever more complex and articulated, in the course of existence and in the course of generations. During their own existence they tend to diversify their organs, adapting them to the surrounding environment, so they tend to develop an increasingly complex structure. Concurrently, over the course of generations, they tend to transmit the acquired characteristics to their descendants, who do the same with their descendants, and so forth. Along these lines, the transformations accumulate, until they overwhelm the initial profile of the organic form. But what provokes the transformation? Why do living beings tend to transform themselves? Because, Lamarck explains, *living beings tend to perfect themselves,* and to this end they tend to diversify their organs and complicate their structure, in conformity with the exigencies dictated by the environment. This does not mean that it is the environment itself that causes the transformation of species. The environment is only the source of stimuli that provoke new needs in living beings, which will continually develop new organs to satisfy their needs. Consequently it is the need that unleashes new actions and habits that then shape the body of the living being, designing organs that it previously did not possess. And yet, if one looks carefully, it is certainly not the environment that causes the transformation of organic forms—*but need is not its ultimate cause either.* In reality, Lamarck observes, living beings feel an unprecedented need arise only because of their innate "tendency" to diversify and complicate their organic form, which coincides with their innate, deep, "tendency" to reach a growing state of perfection (Mayr 1982, 343–62; Pichot 1993, 677; Duris and Gohau 1997, 87). The giraffe, for example, elongates its neck so as to eat leaves on the top of trees, and for this reason, generation after generation, the neck of giraffes tends to get longer. *The need creates the organ,* one would immediately conclude: the need to eat shapes the organ that can satisfy it. Even though widespread, this would not be a correct interpretation of Lamarck, however. The question that is really key for him is another one: why does the giraffe feel the need to eat more food? Because, he answers, by eating more the giraffe will be able to further articulate its organism, it will be able to complicate its own organization, taking it to a higher stage of perfection. This is the true motor and the ultimate cause of organic transformation: the innate tendency to perfect (i.e., to diversify and complicate the organic form). Thus, as Lamarck seeks to illustrate, natural history finally reveals itself as a long race toward perfection, which proceeds from down low to on high, in a cosmos that has not ceased to obey an ancient hierarchy. From infusoria to humans, from the most simple and elementary being to the most complex and organized, the *scala naturae* now traces the unidirectional line of evolution, which proceeds from the most imperfect form to the most perfect. Living beings evolve, by degrees, from lower to higher levels, from infusoria to humans, which now appear as the complementary magnets of evolution, one the etiological magnet, the other the teleological magnet, one the etiological prototype (infusoria), the other the teleological prototype (humans) of natural history. Precisely by virtue of this ingenious combination of etiological prototype and teleological prototype within a compact theoretical discourse, Lamarck succeeds in imagining a rectilinear transformation of species, which does not fall back into the morphogenetic circularity of the morphological prototype (from Diderot to Goethe). It is only thanks to this combination of etiology and teleology that he can conjecture an irreversible transformation of organic forms, hence abandoning the idea of an original Form, cyclically remodeled by all others. As a result, the history of living beings looks like the temporal unfolding of a *scala naturae,* closed at the bottom and the top by its two prototypical forms. And if this is how things stand, it is because this is how it was decreed by the sublime Author of nature. It is because this is the supreme *will* of God, in a universe not too dissimilar from Pico's and Herder's, all stretched out toward its Creator. "By these wise precautions, everything is thus preserved in the established order; the continual changes and renewals which are observed in that order are kept within limits that they cannot pass; all the races of living bodies continue to exist in spite of their variations; none of the progress made towards perfection of organization is lost; what appears to be disorder, confusion, anomaly, incessantly passes again into the general order, and even contributes to it; everywhere and always the will of the Sublime Author of nature and of everything that exists is invariably carried out" (Lamarck 1984, 55).

3.3. What is life? In order to appreciate the novelty of Darwin's response, it might be useful to reflect upon the formula of the *conservatio vitae* that we find in some classics of political philosophy, for example in Hobbes. Here the idea is that life is a struggle for life, but this struggle is not yet thought of as a struggle for the infinite enhancement of life, for a healing of life from its congenital defectiveness. Such a conception will only arise with Schelling, and after him will spread among German biologists who at the beginning of the nineteenth century will be stunned by his *Naturphilosophie.* This is the new "concept of life" whose discovery Schelling will proudly announce. And the question then is precisely: why will this concept appear so novel? Because, at this point, it will no longer be the case that life is moved by the old law of the *conservatio vitae;* instead, life will depend on the new law of the *salus vitae.* In other words, life will no longer express the drive of all living beings toward their own preservation but rather the drive of life itself toward its own redemption and healing. The old statics of the *conservatio vitae,* which always aimed for the defense of the living being and the maintenance of the *status quo,* is thus replaced by the new dynamics of the *salus vitae,* which aims to reinforce life as such and its infinite evolution. The word itself, "life," no longer refers to a single existence, to a single living being, but rather to the abstract force of life that flows in all living creatures. Undeniably, this new connotation of life digs a deep abyss between Hobbes and Schelling, just as it does between Hobbes and Darwin. From a life subdivided into forms-of-life, *heteronomous,* one passes, with Schelling first and Darwin later, to a life detached from every form, *autonomous.* Just as a few years before one had transitioned, with Kant, from a human will that was bound and *heteronomous* to a human will that is absolute and *autonomous.* Along the same lines, from a static diagram of life, nailed to the natural forms that *contain* it, one moves now to a dynamic diagram of life, that forms-of-life cannot restrain but only *convey.* As Darwin says, life looks like a language, of which the various species represent so many dialects (Darwin 2009b, 370–71).

3.3.1. Thus, if Darwin is not the first to speak of a "struggle for life" and the idea surfaces in previous authors, it does however surface there with a different meaning. In this regard, it should also be noted that the very idea of a "war of nature" has a long and complicated history, strictly intertwined with that of the "equilibrium of nature" (La Vergata 1990). Let us give just one example. Outlining his curious political philosophy of nature, Carl von Linné, aka Linnaeus, already describes the world as the theater of a "war of all against all," where "a horrible slaughter" takes place; and already for him this war is to be seen as a conflict between natural species, in the course of which the animals devour the vegetables and each other, without any pity whatsoever. That said, for Linnaeus all of this is part of a more general "economy of nature," in that generation and destruction safeguard the overall equilibrium of nature and the conservation of each species. Generation compensates for losses, destruction compensates for excesses, and both together keep a balance in nature. "The economy of nature consists in *generation, conservation,* and *destruction,* so that the Creator's work may be preserved intact, and all of Nature conspires to this end" (Linnaeus 1767/1770, 18). If this is how matters stand, Linnaeus continues, it is because nature's economy is actually run by a political government that protects nature's "order." This is why he defines nature as a *Respublica naturae,* as a "State." Nature's *oeconomia* presupposes a *politia* in nature. This means that the natural order is kept in balance only thanks to a hierarchical order among the forms-of-life, designed by the Most High, who imposes on some that they be the means of subsistence and conservation of others, and imposes on all that they be the means of subsistence and conservation of the supreme being, the final purpose of all nature: the human being. In this light, the natural order looks like an arrangement, a teleological order, sculpted by the will of the Supreme Author, of the demiurge of the *Respublica naturae,* who placed human beings at the apex of his creation. And nothing changes in the long run. The equilibria between species are not altered. The war of all against all is ultimately a false movement, which hides a fixed, immobile, structure, an order among living beings that reflects the unmoved will of God. The natural world is subdivided and articulated in the crystal of a static Form, that is one with the crystal of a Will, of a divine commandment. "The *politia* of Nature is in force in all of the three kingdoms of Nature. In fact, just as people are not born to be governed by governors, but governors were instead instituted to maintain order among the subjects, so the herbivorous animals, by exacting a cruel tribute, exercise their dominion for the benefit of the vegetables, the carnivores act similarly for the benefit of the herbivores, and among carnivores the large for the small, and human beings (to the extent they are animals) for both the larger and the smaller, but principally for themselves, so that the proportion and the splendor of the *Respublica naturae* is preserved. In their turn, all citizens contribute to the majesty of man, the rational being and lord, whose task is to recognize the Supreme Author of the State" (17).

3.3.2. The same equation between war and equilibrium of nature, which puts the law of the *conservatio vitae* into effect, is made by many other philosophers and naturalists of the period (King, Bonnet, Spallanzani, Bayle, Herder . . .) and even by some of Darwin's contemporaries (De Candolle, Lyell, Blyth . . .). It is also made by Lamarck, just as it is made by Buffon, one of Linnaeus's fiercest opponents. With one significant qualification however: Buffon strongly emphasizes that in the natural order there must be always the same quantity of life, which in his view amounts to the quantity of "organic molecules" that exist in nature and make up those "organized beings," the living, whose form bears the imprint of an "internal matrix." For Buffon, accordingly, life can be *quantified* even before it can be *qualified*. There exists a (quantity of) *life* or a (quantity of) *vitality* that doesn't coincide with the *form-of-life*. But we don't know much else about this (quantity of) life. Nonetheless, it is precisely such a constant quantity of life that finally seems to be the most reliable sign of the natural world's stability, for this quantity remains stable over time, the war of nature notwithstanding, and therefore does nothing but corroborate a static view of the Great

Chain of Being. Life, in this way, is to be quantified, and yet it cannot be really *abstracted away* from quality. Life, for Buffon, is not yet a force that one can completely disentangle from form. The war of nature is there to continuously reapportion the same fixed quantity of organic matter into natural forms. If this quantity were not fixed, Buffon argues, it would be the very Form of nature that would suffer the worst consequences. One should then be able to conceive the inconceivable: the origin of new species, never seen before. "There exists on earth, in the air, in the water a given quantity of organic matter that nothing can destroy; and there exists a given number of matrixes [*moules*] which can absorb it. The matrixes, or individuals, are continuously destroyed and renewed, but they always amount to the same total, and they are always proportional to the quantity of living matter. If the quantity of living matter were to be superabundant, if it were not employed and entirely absorbed by existing matrixes, new matrixes would be brought about and one would see new species appear, because living matter cannot remain inactive, it is always active, and as soon as it is united to organic parts it brings forth organized bodies. The very form of Nature depends on this great combination, on this invariable proportion" (Buffon 1765, IX).

3.3.3. The equation between war and equilibrium of nature, which holds the waves of life within the banks of the *conservatio vitae,* plays an important role also in Thomas Robert Malthus's *Essay on the Principle of Population,* which Darwin read and found quite striking. But what did Malthus write that was so striking? Actually there was only one true novelty in his essay, only one new principle, which wasn't that new after all, but was nonetheless destined to change the overall approach to the question of life in the field of naturalistic investigation. The principle was the one mentioned in the study's title: the principle of "population." Briefly said, Malthus argued that the struggle for life does not take place only between antagonistic forms-of-life, but can also take place within the same species. There is not only an inter-specific conflict therefore, there is also an intra-specific one, which breaks out mostly among human beings. This conflict is provoked by an unchecked growth in the *quantity* of the population, as compared to food resources that remain constant. This situation, so goes Malthus's diagnosis, inevitably produces misery and poverty, thus unleashing a ferocious competition among the members of the same population, because the total amount of the population cannot exceed certain limits, given the scarcity of natural resources. This is why some will have to perish so as to allow others to survive. As everyone can easily see, the old law of the *conservatio vitae* is still in force here. All accounts have to be balanced: nature doesn't tolerate its limits being exceeded. And yet life begins to detach itself from the forms-of-life. Thinking about life no longer means thinking about forms-of-life. Life begins to be abstracted and measured like a quantity, i.e., a population, whose dynamics are not necessarily dependent on this or that quality of living beings. The laws concerning the behavior of this or that species do not coincide with the laws concerning the behavior of populations: the former and the latter refer to different (ontological) dimensions. That said, for Malthus the nightmare to be avoided remained Buffon's: to see the untouchable Form of nature turned upside-down and defaced. "The germs of existence contained in this spot of earth, with ample food, and ample room to expand in, would fill millions of worlds in the course of a few thousand years. Necessity, that imperious all pervading law of nature, restrains them within the prescribed bounds. The race of plants and the race of animals shrink under this great restrictive law" (Malthus 1826, 5).

3.3.4. But would the germs of life, had they been able to develop with no hindrances, really have filled millions of worlds, as Malthus proclaimed? Actually, the question wasn't that wacky for the biology of the early nineteenth century, especially in the German area, and probably for this reason Malthus's essay and his reasoning had such a strong impact on Darwin. In the same years (or months) in which the idea of natural selection is taking shape in his mind, Darwin is in fact getting up to date with the cultural novelties that have hit England in the meantime. He was away for five years, wandering around continents, and when he returns the rising star of English naturalism is Richard Owen, whom he immediately contacts. This is a name to remember, because Owen imports the ideas of Johannes Müller into England, and Darwin then starts to ponder them very carefully (Sloan 1986). Müller was a German biologist coming—like many of his German colleagues, such as Kielmeyer, Reil, Von Baer, Oken, Carus, Tiedemann (Lenoir 1983)—from the stormy season of *Naturphilosophie*. In his works, he argues that there exists in nature a "life-force" (*Lebenskraft*) or an organic force (*organische Kraft*) that can be quantified. The idea was new. Previously one had thought of a circumscribed quantification of vital phenomena, such as the strength of a muscle (Borelli), the speed of blood's circulation (Keill), or the amount of air that enters the lungs (Jurine, Lavoisier), but nobody had ever thought of quantifying life itself, the vital or organic *force*. Müller instead contends that one should. There is a measure of life. For him, this does not mean that one can really measure life in our laboratories. It only means that we should postulate that life is quantifiable. So, why do we need this postulate? Why do we need to imagine that life has a measure? Because, Müller replies, life (vital, organic, force) must obey a law of conservation. Life must be chained to the law of the *conservatio vitae*. This entails that on the one hand we can describe life as a quantity, distinct from quality—each germ of life contains a certain amount of force that can be abstracted away from its own form—while, on the other, the quantity of life always remains caged in its quality—the vital force is exhausted each time in the unfolding of its vital form. In terms that are closer to Müller's formulation: there exists in the world a constant quantity of organic matter that reproduces by means of germs; these germs are organic precisely because they contain a certain quantity of vital (or organic) force; but this force dissipates as the germ "evolves" (develops) until it reaches the complete exhaustion of the vital (or organic) force, once the organism reaches full maturity. At this point the vital (or organic) force is entirely concentrated once more into a new germ, which starts the process again from the beginning, by means of reproduction. A theory of epigenesis is subtly being proposed here, a theory in which one finds strong echoes of Blumenbach and his *Bildungstrieb*. It is like a game of moving volumes, which sometimes expand, dispersing vital force, and sometimes contract, concentrating the entire force in a single point, the germ. It is like an intensity pendulum, a game of systole and diastole. And it is a game that is strictly regulated by a *teleological* process through which the organism's force unfolds itself into the organism's form. Here, organic force and organic form are perfectly coextensive and overlap one another because of the total immanence of the vital force in the vital form. A vital (organic) force without a vital (organic) form does not exist, according to Müller. When it comes to life, force and form are one and the same. But there is an inverse proportion between the intensity (concentration) of force and the complexity (development) of the form. This is the hypothesis that Müller puts forth (and that Owen will endorse). The more the form of an organism "evolves" (develops), the more the organism's force is consumed, until it extinguishes itself completely in death. This kind of theoretical model could be applied to individual organisms in order to explain their development, but one could also extend it to species, so as to explain their extinction (as Müller will do). The only thing this theoretical model could not explain (contrary to Owen's opinion) was the origin, rather than the extinction, of natural species. The law of conservation did indeed allow one to imagine that the organic force of an entire species could *exhaust* itself at a certain point, once that species had attained its most "evolved" (developed) form. And this is why the species would subsequently become extinct. But this law of conservation (law of the inverse proportion between the intensity of the vital force and the complexity of the vital form) in no way allowed one to explain the *emergence* of new organic forms. How do natural species originate? To solve the enigma, one needed to completely detach the vital force from the organic form, one needed to abstract, in a much cleaner and more radical fashion, life's propulsive energy from the teleological, predetermined process of the unfolding of the force into the form. Which explains why Darwin, though intrigued by Müller and Owen, felt the imperative need to learn more by looking elsewhere, perhaps in Malthus's neighborhood, or in areas that were rather embarrassing for someone like him, for a young English naturalist in search of answers. Conquering all hesitation, it was already in those areas that Owen had found inspiration when formulating some of his boldest hypotheses. "Some have supposed that Life was only the consequence of the harmony or consentaneous working of these prime wheels of the Machine, but we are compelled to refer this harmony to a cause which uses it as a means whereby to operate through the whole, to a cause which is independent of the individual parts, since it exists before these parts and consequently before their harmonious co-operation" (Owen 1992, 219).

3.3.5. One can assign these enigmatic words, which Darwin probably heard from Owen's own voice during a lecture, the meaning one prefers. One can interpret them as the reformulation of a teleological view of life against a mechanistic one: a mysterious Intelligence would preexist matter and program its organization. Or, instead, one can read them differently: a mysterious life-force without either form or harmony, in one way or another, would generate forms-of-life and harmony. It is plausible to assume that Owen knew Schelling's thoughts on the matter. Schelling was in fact the first to think that life could be an autonomous force. At the end of the eighteenth century, he had been the one to theorize a "dynamic evolution" of living beings, animated by a formless Absolute, a pure Self, in which Nature and Spirit, life and will, intersect and identify with one another. It is to Schelling that some of the great German scientists of the period, like Carl F. Kielmeyer, looked with deep admiration. In his correspondence with the French naturalist Georges Cuvier, Kielmeyer overtly praises Schelling's genius. As for Cuvier himself, in his last lecture of 1832, he recalls that Schelling was the first to put forward the hypothesis of a fundamental unity of life within the multiplicity of living beings, then promptly adopted by Kielmeyer and others (Richards 2002, 248). Schelling had been the first to advance the idea that "life is not a product or a property of animal life, but rather the opposite: animal matter is a product of life" (Schelling 1927a, 568). He had been the first to speak of a life abstracted away from living beings, nourishing "the perennial evolution, the constant transformation" of species and natural forms (Schelling 1927b, 43). He had been the first to whisper that Life as such could even originate new organic forms, and not only lead them to extinction. So, if all this is true, what kind of kinship can we establish between Schelling the prophet and Darwin the scientist? At first glance none—but the matter is complicated. The penetration of German Romantic biology, imbued with *Naturphilosophie,* into England goes back to the last years of the eighteenth century. Samuel Coleridge, a friend of Erasmus Darwin (Charles's grandfather) and author of a posthumous *Theory of Life* in which the hypnotic melody of Schellingian metaphysics resounds (Coleridge 1848), was one of those who most contributed to building a solid bridge between the two cultures, English and German, upon his return from a visit to Germany. Joseph H. Green, one of the most important figures among English naturalists, almost immediately became part of Coleridge's London circle, and soon afterward started to avidly read Schelling and his pupils (particularly Oken and Carus). Green, in turn, introduced Owen to the arcana of *Naturphilosophie.* As if that were not enough, in the 1810s two stars of British naturalism, John Abernethy and William Lawrence, entered into a harsh debate about the essence of vitality. As foreseeable, the controversy centered on the nature of the vital principle, but the horns of the dilemma were still those that Blumenbach had established. Is that principle simply applied to matter, by the hand of the Almighty, or is it immanent to living matter, like a *Bildungstrieb* internal to the organism? While waiting for an answer, English translations of German essays—literary, philosophical, scientific—started to flood the scene, in the *Foreign Quarterly Review,* the *Edinburgh New Philosophical Journal,* the *British Foreign and Medical Review,* and last but not least in Richard Taylor's Scientific Memoirs, a series of scientific publications that Darwin often consulted in the years from 1836 to 1838, the years when he was sketching the outline of his theory. In brief, England was submerged by a wave of Germanophilia during the first decades of the nineteenth century, and by the mid-1830s it was approaching Germanomania. In this general atmosphere, no wonder that Darwin himself began to learn German in 1836, with the help of his brother Erasmus, just after having returned from his long voyage around the world.

3.3.6. In Darwin's notebooks we find traces of the reflections that would take him, step by step, to his major idea, "natural selection," but one can also find traces of his readings. For example there are traces of Müller, read in the Scientific Memoirs, perhaps at the instigation of Owen. One note on Müller helps us to understand what questions Darwin was asking himself at the end of the 1830s. At a certain point Müller clarifies that "organic force also is increased during the organization of new matter." And this explains why a plant, after having released a seed, also releases others, without immediately withering. But this is a clarification that does not accord with the law of conservation that Müller holds so dear. According to the principle of the *conservatio vitae,* in fact, the overall organic force of nature should always remain constant, passing from organism to organism, from generation to generation. It is precisely for this reason that Müller writes that organic force is *also* increased during the process of organization and, to be precise, is increased by "unknown external sources." The words Darwin underlines are the first ones: "organic force also is." His observation in the margin is the following: "With respect to the non-development of Mollusca, which I have sometimes speculated might be owing to absolute quantity of vitality «in the World»,—*the production of vitality,* as argued by Müller from propagation of infinite numbers of individuals from one, is adverse—" (Darwin 1987, 419). The problem that attracts Darwin's attention is the "production of vitality." Darwin too, for a long time, had thought that the absolute quantity of vitality "in the World" was a constant, but Müller's argument says the opposite, putting his previous hypotheses in doubt and jeopardizing the Müllerian principle of the *conservatio vitae.* Will one then need to imagine that there is a production of vitality "in the World"?

3.3.7. The conjecture wasn't new. Green had already contemplated it. And Green, whose passion for Schelling's "evolutionism" we mentioned above, is not just anyone. In fact Green is one of the first to pass from the *scala naturae,* which had predominated for centuries, to the image of the tree of nature, which would definitely impose itself with Darwin. Nature, Green observes, should not be portrayed as a "ladder" (or a "scale") because "the gradation and evolution of animated nature is not simple and uniform; nature is ever rich, fertile, and varied in act and product:—and we might perhaps venture to symbolize the system of the animal creation as some monarch of the forest, whose roots, firmly planted in a vivifying soil, spread beyond our ken; whose trunk, proudly erected, points its summit to a region of purer light, and whose wide-spreading branches, twigs, sprays, and leaflets, infinitely diversified, manifest the energy of the life within." In this passage from 1828 (cited by Sloan 1992, 36), one should not only emphasize the idea of the tree of life, or the reference to "the energy of the life within," but also and above all the idea of an "infinite" diversification of life/nature. For it is precisely this idea, detaching an infinite force of life from the finitude of the living, that then became crucial for Darwin. This is the idea that Darwin, for example, surreptitiously introduces in his reading of Müller, where he notices that the latter speaks of the "propagation of infinite numbers of individuals from one." As a matter of fact, in Müller's paper, the word "infinite" does not appear. And for many reasons, previously discussed, it *could* not appear. Müller just remarks, while probably scratching his head, that life (or vitality) is a quantity, to be sure, but the exact measure of this quantity escapes us, and so does the secret of life. Since, quite literally, things don't add up in nature.

3.3.8. All of this illustrates, at least partially, the deep embarrassment of the naturalists of the period, and perhaps explains why Darwin now and again began to leaf through essays by authors who were certainly not among his favorites. One of these essays is *On the Kingdoms of Nature* by Carl Gustav Carus, a pupil of Schelling, published in the Scientific Memoirs of 1837. Darwin's notes about it are few but curious. He is particularly struck by one idea: life conceived as a "unity through multiplicity." It is the idea of a life autonomous from living forms, which Carus had inherited from Schelling. "After reading 'Carus on the Kingdoms of Nature, their life & affinity' in *Scientific Memoirs* I can see that perfection may be talked of with respect to life generally.—where [']unity constantly develops multiplicity['] (his definition "constant manifestation of unity through multiplicity" this unity,—this distinctness of laws from rest of) universe «which Carus considers big animal» becomes more developed in higher animals than in vegetables" (Darwin 1987, 269–70). These notes are confused, as are many contained in the notebooks, but one can still derive a couple of reflections from them. The first is that Darwin is struck by the counterposition of the unity of life and the multiplicity of forms in which life is expressed or "manifests" itself. The second is that Darwin now seems to understand that one can speak of the "perfection" of life without considering its various embodiments. Said in a slightly different way, Carus makes him see that life perfects itself as life, according to its own laws, according to the laws of life that should be distinguished from the remaining laws of the universe. This idea did not originate with Carus, as we already know. But what matters most is Darwin's enthusiasm: "Good idea, to show life only laws. like universe." *Good idea*. But what was it exactly? Carus's text, which Darwin underlines and glosses, reads as follows: "As it follows from the foregoing observations that life is not a single isolated reality, we shall be obliged to define it generally as the constant manifestation of an ideal unity through a real multiplicity, that is, the manifestation of an internal principle or law through outward forms." This is Carus's *good idea*. And this is what makes Darwin state, a bit later, that "This paper might be worth consulting, if any Metaphysical speculations are entered in upon life." Then he adds, almost as if not wanting to forget, or perhaps not believing what he was writing, "Namely Carus."

3.3.9. In the light of the above, the temptation is all too great to exaggerate the importance of these few notes and to conclude—as a student of Darwin and Romantic biology has done—that in order to fully understand the genesis of Darwin's theory one actually needs to look at German philosophers, especially those belonging to the most speculative current of "panvitalism," like Carus, instead of those who are closer to the materialist vitalism of a Von Baer or a Müller (Sloan 1986). But is it really possible to establish such a strict connection between Darwin and idealist panvitalists, who first thought of an autonomous and standalone life? From a philological point of view, it would be possible if one were able to know a lot more about Darwin's readings and their effects—that said, it would not be much use. It would be risky to argue for an, even remote, derivation of Darwin's theory from Carus's theory of life (or Schelling's theory). The terminology, the type of reasoning, the very *forma mentis* displayed by these thinkers have nothing in common. Darwin belongs to a different world. More than on the historical and philological level, then, it is on the logical and argumentative level that it can be useful, if not enlightening, to test the resemblance of Darwin's ideas, while still in their infancy, to the ideas of others. In fact a strange affinity seems to exist, thus motivating his attraction for a literature that many in his milieu might have considered mysteriosophical. And yet, it would be misleading to discuss such a palpable affinity in philological terms, because Darwin doesn't derive much from Carus's texts (or other philosophers of nature). Rather—and this is the hypothesis we shall follow—*he seems to repeat the same metaphysical move in his own fashion*. This is the move that underlies Schelling's dynamic evolutionism and remains hidden behind Darwinian evolutionism. That move is among the most ancient and recurring: projecting will into life. Except that at this point it will no longer be the projection of a heteronomous will, which subdivides and wears down life into its multiple vital/final forms, but rather the projection of an autonomous will, which will forever detach life from its natural forms and will make it a *purposiveness without a purpose, a pure form of finality*. This is the move that first Kant, then Fichte, and finally Schelling had made, talking of "reflection," then of "(ex)-change," and finally of "identity" of will and life. And this is the move that Darwin did not copy from anyone but realized in his own fashion, following a brand-new path.

3.3.10. Before pushing further into the meanderings of "natural selection," let us recall the problem that many of Darwin's readers encountered as soon as he made his theory known: the problem of finalism (Gilson 1984). Many were under the impression that Darwin's work covertly injected a principle of Intelligence into the natural kingdom. This is not how things stood, but the mistake is worthy of the greatest attention. Admittedly, Darwin never thought that an *intelligent* will is at work in nature, nor did he see natural processes as the products of a divine Providence; but he did seem to think that a hidden and *blind* will is at work in nature—a will devoid of intelligence and consciousness. What is certain is that he never injected a set of purposes and predetermined directions into nature, a set of final forms coinciding with so many vital forms; nor did he ever inject a principle of Intelligence into nature. He was never enticed by hylozoism nor, obviously, by natural theology. Nevertheless, the problem of finalism gave him trouble, and it is probably by coming to grips with this problem that he finally managed to pull the rabbit out of the hat. Again, there are traces of this in his *Notebooks*, for example where he wonders whether it would be possible to speak of "willing without consciousness," or where he says that one should clarify the meaning of terms like "Reason, Will, Consciousness" (Darwin 1987, 612–13). But there are traces of this also in his son's comments: "One of the greatest services rendered by my father to the study of Natural History is the revival of Teleology" (Darwin 2005, 328). In making this remark Francis Darwin was thinking of the words of one the first and most passionate Darwinists, Thomas Huxley: "The most remarkable service to the philosophy of Biology rendered by M. Darwin is the reconciliation of Teleology and Morphology, and the explanation of the facts of both, which his view offers" (328). Or perhaps, he was thinking of the words with which his father had responded to Asa Gray, thanking him for an article wherein he had been pleased to read: "Let us recognize Darwin's great service to Natural Science in bringing back to it Teleology" (Darwin, Francis, 1959, 232). Darwin's (the father) answer had been unambiguous: "What you say about Teleology pleases me especially, and I do not think anyone else has ever noticed the point" (232). That said, as everyone learns in school, Darwin had mocked the old teleology, the old finalism, associated with both natural theology and panpsychism. There are no purposes in nature: nowadays this is still considered one of his main teachings. So, how could this idea be reconciled with all those glosses and scholia, according to which he had reintroduced teleology into the natural sciences? How was it possible to refute

Lamarck because of his theses on the innate "tendency to further perfection" and to brag about having revived teleology? In order to answer this question, there are two options. The first is that of historical research, possibly enriched by references to Alexander von Humboldt, one of the young Darwin's heroes, and by the study of the Romantic biology of the early nineteenth century, from which he certainly drew—with all the pros and cons that this type of inquiry entails (Richards 1992; Richards 2002). The second is that of philosophical investigation, which allows us to raise our gaze to Darwin's *metaphysical move* and to enter the jungle of "natural selection," where the will reduced to a skeleton, to an empty form, to a purposiveness without a purpose, reveals its ferocious and adamantine face. That Darwin was imbued with German culture is a fact that we can take for granted by now. But this is not enough to make him the architect of the world in which, for almost two centuries, we have been living.

3.4. Darwin's theory is a metaphysical drama in three acts that: (a) separates the force-of-life from the forms-of-life; (b) shatters the Great Chain of Being; (c) replaces the law of the *conservatio vitae* with the law of the *salus vitae*. Darwin's is a "view of life" that one can define as a metaphysical drama first of all because of the thousand-year-old conceptual pillars it ends up demolishing. But there is more—and this explains the metaphysical *pathos* that often exudes from Darwin's texts. The notion of life that Darwin puts forth, when we carefully scrutinize it, looks like a metaphysical anamorphosis of Kant's notion of the will. We see, that is, the same speculative scheme at work in Darwin's theory of life and in Kant's theory of the will. The scheme is that of aseity, putting the Self center stage: just as the will establishes its-Self by willing, so life establishes its-Self by living. Here, it all comes down to a matter of self-reference, of self-institution, of autonomy. "Autonomy" signifies obedience to oneself, obedience to the law (*nomos*) that the Self sets for itself (*autos*). According to Kant, the will only obeys itself, this being the deepest meaning of the "categorical imperative." Likewise, according to Darwin, life only obeys itself, this being the deepest meaning of "natural selection."

3.4.1. What are the load-bearing axes of what Darwin himself, circumspectly, calls the "hypothesis" of natural selection (Darwin 1896, 21)? "The basis of natural selection is simplicity itself—two undeniable facts and an inescapable conclusion: 1. Organisms vary, and these variations are inherited (at least in part) by their offspring. 2. Organisms produce more offspring than can possibly survive. 3. On average, offspring that vary most strongly in directions favored by the environment will survive and propagate. Favorable variation will therefore accumulate in populations by natural selection." (Gould 1977, 11). Clearly this is not the *entirety* of Darwin's theory. Moreover no such thing exists as *one* Darwinian theory; several exist, and they are not always consistent with one another. That said, the three aspects mentioned by Stephen Jay Gould allow one to immediately highlight the fundamental concepts of *variation, adaptation,* and *selection,* which constitute the kernel of the "hypothesis" that made Darwin a historical figure. We will only concern ourselves with these three concepts here, trying to show that they conceal theoretical postulates (variation), crypto-teleological notions (adaptation), and anthropomorphic metaphors (selection), all introduced on the basis of speculative more than observational principles, which might be recapitulated under these headings: the principle of ignorance (variation), the principle of perfectibility (adaptation), and the principle of vitality (selection).

3.4.1.0. In order to clear the field of ambiguities and plunge headlong into the Darwinian vision, one needs to premise that, according to its inventor, the hypothesis of natural selection serves to explain the environmental adaptation and the slow "transmutation" of living forms. This means that, from this perspective, it is natural selection that causes (and explains) the mutable adaptation of living beings to their environmental habitat, and not the other way around. There are various ways to dispel this recurring ambiguity between natural selection and what one might call an environmental selection—an ambiguity that for years will cause Darwin himself to hesitate, and will then recur so frequently that it still resurfaces today in the words of some self-identified Darwinists. From a merely logical point of view, it should be noticed that if environmental adaptation were the cause of natural selection, it would be its final cause, instead of its efficient cause, since any adaptation to the habitat is an outcome, a result of the processes of selection. And it should be added that one would thus transform natural selection into: either an intelligent force, an intelligent principle, which works with a view to certain ends (adaptive, environmental); or a subordinate hypothesis, if not a tautological one, for natural selection would not be the cause but the effect of environmental adaptation; or both. This is why Darwin, in the introductory pages of *The Origin of Species,* distances himself from Étienne Geoffroy Saint-Hilaire, Patrick Matthews, and Robert Chambers, guilty of having adopted a perspective that was topsy-turvy compared to his: instead of explaining adaptation by means of natural selection, they intended to explain evolution by means of environmental adaptation (Darwin 2009b, 6–9). Having said this, there are several other ways to dispel the same misunderstanding, which tends to diminish the significance of Darwin's theory—sometimes with the best of intentions—thus neutralizing, deactivating natural selection (Eldredge 1995), rather than seeing it as an explanatory "principle" (Darwin 1896, 21). From a historical point of view, for instance, it is a fact that in the Darwinian biology of the twentieth century the notion of (selective) *fitness* took priority over the notion of (environmental) *adaptation,* to such an extent that the very idea of adaptation—an idea that today is deemed to be "onerous" and obscure (Williams, George, 1966; Burian 1992; Fox Keller 1992)—was repeatedly called into question and problematized from different, sometimes opposite, perspectives. Or, to give just one more example, one can recall the vicissitudes of the concept of niche in the context of a discipline, ecology, which was literally revolutionized by Darwin's hypothesis, thanks to the work of George F. Hutchinson. What took place here is that one passed from the idea of "environmental niche" (Grinnell) or "functional niche" (Elton) to that of "population niche" (Hutchinson), thereby making the category of *adaptation* to the habitat disappear from the very vocabulary of ecology. In Darwinian ecology, nowadays, "the exploitation of evolutionary opportunities" is no longer described "as a process of adaptation *to* a niche" (Colwell 1992, 241–42). Is this at odds with Darwin's original insight into natural selection? As we shall see in a moment, the answer is no. In the end, all this is perfectly consistent with the kernel of the hypothesis that Darwin found so difficult to articulate.

3.4.1.1. *Variation.* Today we may see it as an "undeniable fact," as Gould writes, but this was not the case in Darwin's times. Even if by variation one only meant the "transmutation" of organic forms in the course of generations, this was already a hypothesis and not a crude fact. It was the hypothesis endorsed for instance by Lamarck, who indeed believed that physiological changes acquired over the life of an organism may be transmitted to offspring. Yet, Darwin meant something different by variation. He meant a casual and spontaneous variation, an unforeseeable and unknown variability of the hereditary traits, a sort of inexhaustible *creativity of life.* Darwin did not think that living beings always leave the following generation an identical form-of-life, an identical program of life, as the fixists believed. Nor did he simply imagine that forms-of-life slightly change in the course of their existence and then transmit their acquired characteristics and altered form to the following generation—"soft inheritance," from Buffon to Lamarck, *via* Maupertuis and Diderot. His hypothesis was much more innovative and radical. He dared conjecture that forms-of-life can change over time without any external cause intervening. He dared conjecture that organic forms can alter themselves *by themselves,* from generation to generation. He dared conjecture that during the process of reproduction different forms-of-life spontaneously arise, even if their difference from the original ones is ever so small. Why? Due to which factors? This was the question that at this point everyone, including Darwin, had to answer. In that regard, Darwin's attitude was laconic: we don't know. There were in fact no other answers to be given in the years *The Origin* was being drafted. So, the problem of "variation" was to remain one of his worries until his death. Casual variation had to be postulated in order for his theory of evolution to have both meaning and strength; it needed to be postulated because it is precisely on its basis that the natural selection of some, favorable, variants, to the exclusion of others that are less so, occurs. Once we have made such an assumption, we can speak of "hard inheritance," in other words of a transmission of characteristics that vary *on their own* and then pass through the sieve of natural selection. Nonetheless, in Darwin's times genetics was a discipline that was barely starting. Just as the "modern evolutionary synthesis," combining Darwin's hypothesis with Gregor Mendel's theory and the scientific scrutiny of genetic variation (the so-called new synthesis by Dobzhansky, Huxley, Mayr, Simpson, Rensch, Stebbins . . .), was still a long way off. Halfway through the nineteenth century, free, spontaneous, stochastic variation was nothing more than a fantastic idea, a speculative hypothesis that could be formulated only by appealing to a mysterious *principle of ignorance,* which surgically separated clear scientific observation of the facts from an interpretation, somehow metaphysical, of those same facts. "Our ignorance of the laws of variation is profound. Not in one case out of a hundred can we pretend to assign any reason why this or that part differs, more or less, from the same part in the parents" (Darwin 2009b, 131). This famous confession explains why Darwin, the unwilling metaphysician, constantly swayed between the two hypotheses, hard inheritance and soft inheritance, almost as if wanting to hide what was quite obvious, and yet hard to swallow in those days, namely that he was proposing a definition of life abstracted away from forms-of-life: *life is an unforeseeable variation of forms-of-life, vital processes are casual and spontaneous processes of variation that continually modify living forms and contribute to explain their origin.* This was Darwin's thesis, just as, somewhat earlier, it had been Schelling's. "Life is not a product or a property of animal life, but rather the opposite: animal matter is a product of life" (Schelling 1927a, 568). Life varies *by itself,* Darwin now argues, life is an unknown variable that comes before its variants and induces them to vary ceaselessly. Life is what is underneath, the *subjectum,* the *hypokeimenon,* the fluctuating variation at the base of the mutable shapes of living beings. This is how Darwin abstracts the life-force from its many forms, making its *vitality* coincide, first of all, with its *variability.*

3.4.1.1.1. It is worth noticing that Darwin himself will try to backtrack and domesticate his fantasies in a work entitled *Variation of Animals and Plants under Domestication,* published nine years after *The Origin.* In that book, he will propose the unfortunate theory of "pangenesis," which represents a dramatic return to the idea of soft inheritance (of acquired characteristics). The irony is that two of his most lucid followers, August Weismann and George Romanes, will soon prove that the principle of natural selection is truly significant if and only if one accepts the premise of hard inheritance, and the ensuing assumption of casual and spontaneous variation. For a man as scrupulous as Darwin, in any case, the problem could not be definitely solved. Those were not times in which one could talk about "chance" or stochastic variation, as Jacques Monod and most contemporary biologists would a century later. Each time he is forced to speak of casual variation, Darwin does so regretfully, with a thousand qualifications, convinced as he is that to say "chance" is only a more elegant way of saying "ignorance." "I have hitherto sometimes spoken as if the variations [. . .] were due to chance. This, of course, is a wholly incorrect expression, but it serves to acknowledge plainly our ignorance of the cause of each particular variation" (Darwin 2009b, 106).

3.4.1.1.2. In addition to this, it is worth noticing that Mendel, who first studied the laws of inheritance, had been a pupil of the naturalist Franz Unger. This is curious because Unger had already arrived at the idea of evolution, as Darwin reminds us (xx), and he had already reached the conclusion that the source of variations was hidden within organisms, and living forms varied *on their own.* So much so that Mendel later stated that it was precisely Unger's original conjectures that persuaded him to study the laws of variation (Mayr 1982, 391). But Unger himself, this botanist who would end up in the cellars of history—who had he been inspired by in conceiving the evolution of organic forms? In all likelihood he, too, had been inspired by German *Naturphilosophie,* not along the Schelling–Carus axis this time, but along the Schelling–Oken axis (Gliboff 1998). As a young man Unger had in fact spent some time in Jena, where he had been particularly struck by Lorenz Oken, a product of the Schelling school. It is not so illogical therefore to think that this is where he got his intuition not only that life evolves, but that life varies *by itself,* and thus evolves, producing ever new species. Whatever the case, as Unger will shortly declare, "the cause for the diversity of plant life cannot be an external one but must be internal . . . in one word, each newly originating species of plants . . . must originate from another one" (quoted by Mayr 1982, 391).

3.4.1.2. *Adaptation.* This is the second pillar of Darwinian evolutionism, which is fought about to this day. What does the expression "survival of the fittest"—which Darwin borrows from Herbert Spencer—mean? Broadly speaking, it means that among the varied forms-of-life, only some will reproduce and survive, only those that prove to be the "fittest" to their conditions of life and to the "struggle for life" that they must engage in willy-nilly. But let us be more specific. What does the word "fittest" mean in this context? Strictly speaking, it cannot mean most adapted to a specific purpose, to a predetermined end, because in this case we would fall back into a teleological vision of evolutionary processes, thus making living forms into final forms whose "transmutation" would regularly aim to reach a functional, intelligible, and predetermined *goal* (vision in the case of eyes, prehensile operations in the case of hands, and so on). In Darwin's view, on the contrary, the evolution of species is not directed by any *vis a fronte,* by any (more or less intelligent) force that would orient and prefigure the transformation of species. The only active force is a *vis a tergo,* the generic, nonspecific force of life as such, which pulsates in all living beings and makes them fight for survival, for their bare life. Is this a thesis that deletes all kinds of teleology from nature? In reality, it is not. This is a thesis that, despite first appearances, still injects a strange purposiveness into nature. As a matter of fact, only the competition among forms-of-life will tell us which is best adapted and will survive. This competition will not say which of these forms is best adapted to a particular purpose, it will rather tell us which is best adapted *to live (and therefore to reproduce).* It will not say which living beings are best adapted to a certain form-of-life, but rather which living beings are best adapted *to life (and therefore to reproduction)*. Now, if this is no longer a teleology of the forms-of-life as one had thought of it in the past, it is because, if one looks carefully into it, this is a teleology whose aim is Life (with a capital L). It is the unprecedented teleology of life that Kant had already imagined when, in the third *Critique*, he had advanced the idea of a "purposiveness without a purpose." By adopting Spencer's category of "the survival of the fittest," Darwin repeats the same speculative move. In his perspective, those who survive are the best adapted, but not to a preestablished form-of-life, to a final form; rather, they are the best adapted to life itself, understood as a purposiveness without a purpose. "There can be purposiveness without a purpose, insofar as we do not posit the causes of this form in a will, and yet can grasp the explanation of its possibility only by deriving it from a will" (Kant 1987, 220). According to Kant, as we have seen before, one can speak of a purposiveness without a purpose each time that one cannot assign the cause of a form to a *specific volition*, and yet one can conceive that form only by deriving it from a *generic will*, which cannot be associated with any preestablished purpose. This means, returning to Darwin, that the "fittest" survives not because it wants *this or that* form-of-life more than others, but because it wants life *in general* more than others. Not because it expresses, more than others, the *specific volition* for a form-of-life, but because, more than others, it expresses a *generic will* for life. The fittest *wanted to live*, struggled for life, won the battle for life, and if they succeeded, ultimately it is because they managed to "better" express such a pure will to life, such a peremptory will to survive, *which is life itself, which is life as such, an abstract and generic life, governed by a "purposiveness without a purpose."* If we keep this speculative model in mind, Darwin's "one long argument" becomes, perhaps, less opaque. It is true that he reintroduces teleology into biology, as his son would later note, but he reintroduces it in a very unusual manner. To be precise, he reintroduces it each time that he resorts to the words "best" or "better," both a blessing and a curse of Darwin's thought. The "best" or the "better adapted" are in fact qualifications that establish a rank order of merit for the forms-of-life, a ladder of greater or lesser "perfection." However, this perfection should not be measured against a preestablished purpose, as when we observe that a shoe fits us perfectly, or better than another one does. Rather, in this case too, one needs to think back to Kant and to his distinction between an "internal" and an "external" perfection. The perfection of the shoe is external perfection, namely "utility," since the shoe finds its purpose *outside* of itself. The form-of-life's perfection is instead internal perfection, since it has its purpose *within* itself. Equally, Darwin says, one form-of-life reveals itself as better than another, or "better adapted" to survive, not because it responds better to an *external purpose*, but because in it life itself pulsated "better" (or more). It is the force-of-life within the form-of-life that produces its *internal* perfection and makes it "better" than others. Seen in this light, which are the species that evolve and survive, being "better adapted"? *They are the ones that vary the most, Darwin replies, they are those that prove to be "best adapted" to life (and reproduction), proving themselves more adaptable, and more variable.* The more they vary, thus dynamically adapting to "the struggle for life," the more the forms-of-life turn out perfect, adapted, and advantaged. This is what "fitness" means, and, above all, what "perfection" means in Darwin's natural world: not the motionless perfection of forms, but rather the ever-moving and ever-changing perfectibility of an organism that wants to live (and reproduce) as much as possible, feeling that life pulsates in its-Self as a force that is both internal and propulsive. Kant too had already spoken of this weird perfection of the living being, which does not so much consist in a definitive fulfillment of its-Self, as in a perennial reconfiguration and redefinition of its-Self. For Kant, living beings are "perfect" in that they are their own purposes, which can never be achieved but nonetheless, precisely for this reason, keep the organisms alive. Similarly, Darwin states, the perfection (and adaptation) of the forms-of-life should be measured against their internal perfectibility. The more the species is perfectible and adaptable, in other words, the more it tends to vary, and the more it becomes perfect and fit for life as such. As a result, unsurprisingly, the moral of Darwin's autonomous life comes to prolong the moral of Kant's autonomous will: "perfection" and "perfectibility" become interchangeable terms. Just as for Kant it was only a matter of *enhancing* the will itself as much as possible, under the aegis of the categorical imperative, so for Darwin it is only a matter of *enhancing* life itself as much as possible, under the aegis of natural selection. In both cases, autonomy (of life or will) flows into a metaphysics of *empowerment* (of life or will) whose gradients and differential intensities it will now be necessary to specify. As Darwin himself emphasizes: "On this same view we can understand how it is that in each region where many species of a genus have been produced, and where they now flourish, these same species should present many varieties; for where the manufactory of species has been active, we might expect, as a general rule, to find it still in action" (Darwin 2009b, 413). Paraphrasing, wherever life prospers with vigor and strength, life proves itself to be more variable. And in order to vary life basically needs two things. First, it needs the *will to live*. Life grows stronger where the desire for life also does. "Look at the most vigorous species; by as much as it swarms in numbers, by so much will its tendency to increase be still further increased" (Darwin 2009b, 53). Second, it needs to *perfect itself*. Life reigns where an infinite perfectibility reigns. "Natural selection will not necessarily lead to absolute perfection; nor, as far as we can judge, can absolute perfection be everywhere predicated" (166).

3.4.1.2.1. These are not merely stylistic expedients, as one might think. Quite the contrary, we find ourselves at the very center of Darwin's argument, of his grandiose "view." The perfectibility principle, for example, splits into two subordinate principles, which we could call the *principle of diversification* and the *principle of specialization*. Both are cardinal principles of Darwin's theory. The diversification principle prolongs and qualifies the perfectibility principle, postulating that species tend to differentiate themselves as much as possible: the more they branch out, detaching themselves from one another in the great "tree of life," the more they increase their chances of survival as well as their vital potentials. "The green and budding twigs may represent existing species; and those produced during each former year may represent the long succession of extinct species. At each period of growth all the growing twigs have tried to branch out on all sides, and to overtop and kill the surrounding twigs and branches, in the same manner as species and groups of species have tried to overmaster other species in the great battle for life" (104). The specialization principle prolongs and qualifies the perfectibility principle on a different level, that is the level of the organism: the more organisms specialize their organs, modifying them with respect to one another, the more their chances of survival increase, as do their vital potentials. "The truth of the principle, that the greatest amount of life can be supported by great diversification of structure, is seen under many natural circumstances. [. . .] As a general rule, the more diversified in structure the descendants from any one species can be rendered, the more places they will be enabled to seize on, and the more their modified progeny will increase" (88, 92). In both cases the forms-of-life, whether species or organisms, increase their vital potentials and *strengthen their life-force* in proportion to their predisposition to change and *deform their form-of-life.* Thus, on the one hand, a species is so much stronger than its specific identity the more it tends to vary and to *re-speciate* itself (diversification of species). While, on the other, an organism is so much stronger than its organic identity the more it tends to vary and to *re-organize* itself (specialization of organs). On both sides, species and organism, the form-of-life appears all the more "perfect" the more its profile appears to be "perfectible," the more it appears to be 'formable' and deformable by the force-of-life. And on both sides a *principle of localization* of variability (geographic or structural) remains in force, by virtue of which the more a characteristic (specific or organic) varies and the more it *tends* to vary. As a result, life begins to exhibit *zones of intensity,* which the naturalist needs to do no more than touch. Life begins to distribute itself into fields in which its force, its heat, is concentrated. Life, now here, now there, seems to be set on fire. "It is inexplicable on the theory of creation why a part developed in a very unusual manner in one species alone of a genus, and therefore, as we may naturally infer, of great importance to the species, should be eminently liable to variation; but, on our view, this part has undergone, since the several species branched off from a common progenitor, an unusual amount of variability and modification, and therefore we might expect the part generally to be still variable" (416).

3.4.1.2.2. "We can see why throughout nature the same general end is gained by an almost infinite diversity of means" (466). This is Darwin's crypto-teleology, or his "teleonomy" (Mayr 1961; Monod 1971), that displays an unwitting Kantian matrix. Deprived of such an invisible support, Darwin's "hypothesis" would immediately collapse. This is one of its metaphysical—rather than scientific—nerves. And precisely because of that, it is not surprising that the censorship of his most devoted advocates has, with the passing of time, repeatedly focused on this feature. Contemporary Darwinists argue in unison that there is no trace of finalism or teleology in the founding father's texts. One can at most speak of "teleonomy," which is something quite different. Now, in point of fact, it is true that there are two different types of teleology: the pre-Darwinian teleology, that of final/vital forms, and the Darwinian teleology ("teleonomy"), that of "purposiveness without a purpose." Just as there exist two different views of life: the pre-Darwinian view of life, subdivided into its natural forms, *heteronomous*, and the Darwinian view of life, abstracted away from natural forms, *autonomous*. Just as there exist, as seen before, two different views of the human will: the pre-Kantian view of the will, subdivided into its multiple volitions, *heteronomous*, and the Kantian view of the will, abstracted away from volitions, *autonomous*. Just as there exists, on a further level, a link between the two worlds, the Kantian world of the autonomous will and the Darwinian world of autonomous life, which compose one single world, ours, based on the idea of a power, of a *dynamics* of the living and willing Self (*aseitas*). In the light of this, however, it is quite evident that there remain abundant traces of teleology in Darwin's texts: we only need to understand which kind. At the same time, it is no wonder that today's biologists tend to deny Darwin ever advanced a teleological view of life: again we only need to understand which kind. To put it in a slightly different way, for decades all Darwinists—just like Darwin himself—have made use of terms that are highly ambiguous and questionable: terms like "better" or "best." That said, they are certainly right to maintain that Darwin jettisons the old teleology of vital/final forms, instead of appropriating it, as some persist in thinking (Gilson 1984), since Darwin glimpses a different teleology in natural processes, a teleology that permeates all Darwinian science and thus creates problems from which this science is periodically forced to extricate itself. Two quotes can testify to this. The two voices speaking are: first, that of a heterodox representative of Darwinism, then that of an orthodox one. "Darwin maintained that evolution has no direction; it does not lead inevitably to higher things. Organisms become *better* adapted to their local environments, and that is all" (Gould 1977, 13, my italics). Let us suppose we agree, with reservations, that for Darwin organisms adapt better to their environments. If this were the case, shouldn't we admit that evolution has a direction, dictated by the environment? Furthermore, if an organism is "better adapted" to the habitat, wouldn't it be a "higher" form of life? Let us say then, changing tack, that for Darwin natural selection leads "[. . .] to survival and the maintenance or *improvement* of adaptedness" (Mayr 1991, 164, emphasis added). In this case, too, what will the criterion of "improvement" be? On what basis will we be able to tell that one form-of-life is "improved" and "better" than another? What is the secret of natural selection?

3.4.1.3. *Selection.* This is the keystone to Darwin's theory. In *The Origin* there are multiple definitions for it. "If variations useful to any organic being ever do occur, assuredly individuals thus characterized will have the best chance of being preserved in the struggle for life; and from the strong principle of inheritance, these will tend to produce offspring similarly characterised. This principle of preservation, or the survival of the fittest, I have called Natural Selection. It leads to the improvement of each creature in relation to its organic and inorganic conditions of life; and consequently, in most cases, to what must be regarded as an advance in organization" (Darwin 2009b, 102–3). Darwin, like other innovators, sometimes finds it difficult to choose the right words. Moreover, he finds it difficult to keep his eyes fixed on the blinding sun of his "hypothesis." When, rather than theorizing, he recapitulates his ideas, he often tends to decrease the emphasis on living beings' profound *fitness* for life (reproduction) by insisting on their superficial *adaptation* to the conditions of life (environment), which is only a side effect (and remains a concept that cannot be easily situated in the context of his theory). That said, the whole of his "one long argument" tends to dramatically diminish the importance of the "conditions of life" in which, traditionally, naturalists perceived living beings' "final causes." "Natural history has, moreover, a principle which is peculiar to it [. . .]; it is that of *the conditions of life,* commonly termed *final causes.* As nothing can exist without the concurrence of those conditions which render its existence possible, the component parts of each being must be so arranged as to render the whole living being possible, not only with regard to itself, but also with regards to its surrounding environment" (Cuvier 1817, 3). From this perspective—which was the one that Geoffroy Saint-Hilaire, Matthews, and Chambers had parsed in evolutionary terms, the same one that Darwin criticizes in the opening pages of *The Origin*—the "conditions of life," dictated by the environment, exercise an *active* pressure, instead of a *passive* pressure, on the transmutation of living beings; they express a force that orients evolutionary processes, instead of a mere obstacle against which, each time, those processes impact; ergo, from this point of view, the "conditions of life" represent positively determining factors, not negatively resistant ones. And this was precisely the old, backward teleology of life forms sculpted by "final causes," from which Darwin wanted to distance himself, in spite of many ambiguities and vacillations. It was from an altogether different idea, that of an increasing adaptation of the living being to life and reproduction (*fitness* rather than *adaptation* in the jargon of later Darwinism) that he had taken his cue several years earlier. Just as it was by pondering this very idea that he had also concluded that there exists a second "principle" of evolution parallel to natural selection, which he defines as "sexual selection." "The structure of each organism is chiefly adapted to the sustension of its life, when full-grown, when it has to feed itself and propagate" (Darwin 1909, 42).

3.4.1.3.1. At the level of language, where all these difficulties come to light, Darwin often yields to the lexicon of the *conservatio vitae,* even though he is opening up a new horizon, the metaphysical horizon of the *salus vitae,* which prescribes an "improvement" and an "advance in organization" of all forms-of-life, instead of their simple preservation. And he does even worse when he writes, more than once, that natural selection skims the variations that prove to be most "useful" to living beings in the eternal battle for life. The term "useful," in fact, introduces a principle that should no longer be presupposed in his discourse, and which ends up undermining its coherence. Let us call it the *principle of utility.* If this principle were valid, Darwin's theory would ultimately be a naturalistic version of utilitarianism, according to which every living being in nature acts according to its "utility," in a manner similar to what individuals do in society. What is more, one would need to postulate that natural selection possesses a comprehension, however minimal, of this "utility," on whose basis it performs its choices, one after the other. As a result, one would fall into a sort of providentialism, and one would erroneously deify evolutionary processes. For one would inject an occult comprehension—of what is "useful" to the living being—into nature. It is Darwin himself who complains about this wrong interpretation of his theory. "It has been said that I speak of natural selection as an active power or Deity . . . " (Darwin 2009b, 63). Such interpretation is mistaken, Darwin objects, because one should not confuse an anonymous and impersonal regularity, a natural law, a "principle" that is at work in evolutionary processes, with a personified power that orients them in a conscious and intelligent fashion. But where do the roots of the mistake lie? Actually, they lie in Darwin's own words, which surreptitiously introduce a principle of utility, where one should instead talk of a *principle of vitality.* This is the sole and exclusive principle for selection: not the "useful" but the vital. In brief, what is the criterion according to which natural selection operates, chooses, selects? *It is not the nuanced criterion of the more or less useful, but rather that of a very sharp alternative: life or death.* The variations that are successful, and that are thus selected, are not variations that are *relatively useful* to the organism, they are variations that are *absolutely vital* to its survival. The variations are either for life or for death. And this is all natural selection sees. Therefore, there is no need to speak of "utility". All that natural selection needs, in order to be operative, is to keep on choosing, to keep on selecting. Natural selection, in this sense, is the act of a will. But this will, anonymous and impersonal, is never combined with the comprehension of a purpose: the "useful." Rather, it is nailed to its condition of possibility: vitality.

3.4.1.3.2. For the sake of simplicity, we could say that natural selection performs its choices in such a way as to be able to perform further choices in the future—this is its uniform and regular criterion. It chooses between alternate, competitive, demands for life and it selects in such a manner that there will be ever more life on which to operate its selection. In this perspective, natural selection comes off as a reiterated "yes" to life. It is the "yes" life says to itself, choosing itself, selecting itself, in order to be able to do so tomorrow as well. It is a "yes" that is not addressed to the "utility" of individual living beings, which is invisible to life as such, but to the vitality of their emerging characteristics, in which a pure will to life, a demand for life, a will to life of life itself, takes shape. In natural selection, life's vitality thus comes to coincide with the simple willfulness of life, a willfulness devoid of preestablished and "useful" purposes, a willfulness that precedes any utilitarian volition. Selection does nothing more than grope for the pulsations of the greatest vital potentialities, which are reproductive potentialities, and it does so with blindfolded eyes, without knowing beforehand where it will end up. It is only on the level of forms-of-life that one can pose queries regarding utility; it is only in the eyes of a more-or-less intelligent living being that something can appear to be "useful." While on the level of bare life and pure vitality, which is natural selection's level, none of this counts. Selection is insensitive to utility: it is sensitive to vitality (reproduction). Selection is nothing but the act by which life inflates its-Self and cries out its will to life. It is a metaphysical drama that Darwin actually puts on stage. It is a passionate and exasperated description of natural processes that are reread, always, in *vital* terms. It is a theatricalization that places the "battle for life," the "struggle for life" at the center of the natural scene, where it is always a matter of *life or death* for the forms-of-life subject to variation. And natural selection does nothing else: *it decides for life or death. Life, that is, decides to live in Darwin's view, just as the will decides to will in Kant's view.* With this decision, the autonomous will crosses over into autonomous life. The purposiveness without a purpose of the Kantian will becomes the purposiveness without a purpose of Darwinian life—under the aegis of the pure willfulness of the will (autonomy of the will) in the first case, and of the pure willfulness of life (autonomy of life) in the second. At that point, the enigma of aseity penetrates into nature. This is not the enigma that human beings had faced up to that moment, the enigma of Deity. For the Deity of life is a Deity that, in the end, *does not even know it is such.* Darwin, even though with some effort, makes this point: natural selection is an *unconscious* selection. It is a specification whose importance one should not underrate.

3.4.1.3.3. The overarching architecture of *The Origin* is quite interesting. The first two chapters are devoted, respectively, to the *Variation under Domestication* and the *Variation under Nature.* As is well known, here (and in the following two chapters) Darwin draws a parallel between "artificial selection" and "natural selection." Just as breeders, Darwin states, select the "best" varieties of plants and animals under domestication, nature analogously selects the "best" varieties of plants and animals in their wild state. The aspect Darwin incisively insists on, however, is the "unconscious" aspect of this selection, both artificial and natural. This is the aspect of his theory that really, honestly, gives him the impression of not deifying nature. What does it mean, from Darwin's point of view, that selection is unconscious? In the case of artificial selection it means that breeders occasionally *don't know* what variety they are trying to obtain, killing certain individuals while letting others breed. Their selection is part methodical and part unconscious; it is a selection sometimes oriented toward a preestablished and precise purpose, of which they are therefore conscious, and sometimes not. "At the present time, eminent breeders try by methodical selection, with a distinct object in view, to make a new strain or sub-breed, superior to anything of the kind in the country. But, for our purpose, a form of Selection, which may be called Unconscious, and which results from everyone trying to possess and breed from the best individual animals, is more important" (25). So, if methodical selection is distinguished by the knowledge of the purpose one is consciously pursuing, unconscious selection is distinguished by the ignorance of the purpose one is trying to pursue. In unconscious selection a will chooses and selects, of course, but without eventuating in any clear and distinct volition, in any utilitarian volition. In unconscious selection a will is at work, but remains so in its pure and abstract willfulness, detached from all conscious and concrete volition. Breeders' unconscious selection is prompted by the desire for the "best" individual animals. Yet this desire doesn't follow any predetermined direction. For this reason, Darwin argues, artificial selection can be equated with natural selection, which is neither oriented nor directed. Natural selection is an unconscious selection, that is a blind selection, which doesn't pursue purposes, methodically, but rather decides, unconsciously, for life at all costs. This entails that natural selection operates not on the basis of a purpose, as represented by the utility of the living being, but rather on the basis of a purposiveness without a purpose, that of life itself: a purposiveness that comes to coincide with the absolute, autonomous, pure willfulness of life. First, life is a matter of *absolute* willfulness, for Darwin, because life is detached from (the comprehension of) all purposes, from (the comprehension of) the results that are gradually obtained. Second, life is a matter of *autonomous,* not heteronomous, willfulness because it expresses the empty subjective form of the will, not the full objective matter of the purpose. Third, life gives body to a *pure* will that, being devoid of specific contents, cannot but be "unconscious." Thus, in the absolute, autonomous, pure willfulness of life the three basic principles of Darwinian thought merge, *the principle of variability (variation), the principle of perfectibility (adaptation), the principle of vitality or unconscious aseity of life (selection).* These mark the three stages by which Darwin (a) separates life from all forms-of-life, (b) shatters the Great Chain of Being, and (c) replaces the scheme of the *conservatio vitae* with the scheme of the *salus vitae*. Such is his revolutionary vision of life—and surely, *"There is grandeur in this view of life"* (429, emphasis added).

3.4.1.4. This long metaphysical formula—combining variability, perfectibility, and the unconscious aseity of life—allows us to shed light on some further aspects of Darwin's theory. To begin with, consider again the *principle of ignorance*. Darwin mostly resorts to it when he discusses variation. In that context the principle has a purely methodological meaning: we know nothing about the laws of variation, and yet we can speak of a casual and spontaneous variation. But let us now think about how unconscious selection operates. In this case the principle of ignorance loses all methodological sense and acquires a purely metaphysical sense: natural selection doesn't know what the fruits of its operations will be, and yet it is operative. In fact, it remains operative precisely by virtue of its ignorance (otherwise it would turn into an intelligent principle). Here, it is as if the two levels of Darwin's reasoning, the methodological and the conjectural, the metatheoretical and the theoretical, were superimposed on one another. The principle of ignorance belongs to both registers of enunciation, and rotating around it, it is as though the principle of unconscious aseity of life (selection) were reflected back onto the principle of variability (variation). Thanks to the principle of ignorance, the two other principles end up expressing one and the same idea: the abstraction of life as a *blind* force, devoid of directionality, which establishes itself both in the *blind* detachment of life from all its natural forms (variation) and in the *blind* beat of life in all its transmutations (selection). We have passed from the principle of variability to the principle of the unconscious aseity of life, but we could have followed the opposite path, for the principle of ignorance is like a bridge that brings us from one extreme to the other of the same metaphysical equation, joining the methodological and the conjectural levels of Darwin's argument into one single, and highly speculative, Möbius strip.

3.4.1.5. The principle of ignorance thus links the principle of variability and the principle of unconscious aseity of life, which are behind the two basic axioms of Darwin's theory: the axiom of variation and the axiom of selection. The axiom of variation tells us that vitality consists in the absolute variability of life. The axiom of selection tells us that vitality consists in the absolute willfulness of life. Now let us make the two axioms flow into one another, using the principle of ignorance, which enables us to go from one to the other. What happens then? It happens that we can be affected by a kind of strabismus: not only do we see variability as a condition of possibility of selection, but selection itself is now seen as a condition of possibility of variability, by simply inverting the direction of the argument. Hence the idea, explicitly articulated by Darwin, that selection can actively *induce* variation. "Those points in our domestic animals, which at the present time are undergoing rapid change by continued selection, are also eminently liable to variation" (Darwin 2009b, 120). Perhaps, this is the reason Darwin, inadvertently, tends to define only those variations that are not subject to the pressures of natural selection as "spontaneous"—almost as if the variations subject to selection were not as "spontaneous" as those not subject to selection, almost as if they were *induced* by selection, unlike those that are truly "spontaneous" (171–76). In any case, first on the level of reasoning, then on the level of definitions, in Darwin's "one long argument" the axiom of variation and the axiom of selection seem to pass into one another, through that metaphysical bridge that is represented by the principle of ignorance.

3.4.1.6. Typical of Darwin's approach is his insistence on the principle of continuity, which seemingly revives the idea of the Great Chain of Being. As he often reminds us, "*Natura non facit saltum*" (414, emphasis added). But what does Darwin mean by continuity? Surely, he means, as his predecessors did, the continuity of the natural kingdom. For centuries, people had believed that the forms-of-life are placed in a *scala naturae* along which each living being differs only minimally from the next one. And this is what leads Darwin to the (well-known) thesis of evolutionary gradualism. Yet, in his eyes this thesis should be connected to another, a less traditional one, that is species nominalism. Species are only names or labels that serve to group individuals that, nonetheless, are never identical: the differences between individuals of the same species are certainly less pronounced than those existing between individuals of different species, but such differences do exist. And it is thanks to them that we can think of a *gradual* transformation of species (one gradually changing into another). All "these differences blend into each other by an insensible series" (41). Thus, the principle of continuity remains in force for Darwin, who, rather than refusing it, seems to radicalize it. Where is the novelty, then? It lies in the fact that a principle of continuity that is radicalized to such an extent becomes incompatible with the *principle of the fullness of being*, which is the real keystone of the Great Chain of Being. One might say that two continuities exist: on the one hand a superficial and closed continuity, on the other a deep and open continuity. The superficial and closed continuity is that of the old metaphysical world, wherein natural species were substantial forms, immobile, intangible, gathered beneath the gaze of an Intelligence who embraced the entire cosmos. In a world like this, the principle of the fullness of being applies because the universe is a space subdivided into stable forms, between which there exists no empty place. There is no void between one step and the other of the *scala naturae*, and there is no void externally; there is no strip of reality outside natural forms. The deep and open continuity is instead that of the new metaphysical world, wherein species don't have any real, ontological texture because continuity is established at a deeper, individual level. In this world, the statics of the old natural kingdom are replaced by the evolutionary dynamics of life, which decrees the mobile "perfectibility" of all living beings. As a consequence, the principle of the fullness of being no longer applies. Here, one should rather speak of an intrinsic emptiness of being. For being is life and life never lives enough. Life is a *will to live.*

3.4.1.7. For Schelling, as we have seen above, the basic properties of life were three: defectiveness of life, progressiveness of life, priority of life with respect to the forms-of-life. These properties now return in Darwin, as is proved by his reflections on the hypothesis of an originary "prototype" of all life forms. The hypothesis that fascinates him is not that of a teleological prototype, which would make no sense in a theoretical perspective like his, nor that of a morphological prototype. Darwin is rather tempted by the hypothesis of an etiological prototype, by the idea of a primitive form-of-life, or of a "common ancestor," from which all ensuing forms-of-life originated. If we take the idea of a gradual diversification of species on the "tree of life" seriously, then the hypothesis of an etiological prototype logically follows. And Darwin does indeed wonder about it. "I believe that animals are descended from at most only four or five progenitors, and plants from an equal or lesser number. Analogy would lead me one step further, namely, to the belief that all animals and plants are descended from some one prototype." However, Darwin immediately adds, "analogy may be a deceitful guide" in this case (424). The gradual transformation of species can in fact lead to a situation in which no structural analogy persists between the original species and the derived one. It is possible for a descendant to have been "so much modified as to have lost all traces of its parentage" (370). We are not sure of that, but we cannot exclude this possibility either. In point of fact, no one can tell, since the fossils at our disposal do not allow us to even vaguely outline the species' entire genealogical tree, up to the (conjectured) common progenitor. Therefore, Darwin argues, one cannot exclude the possibility that there are "many different forms" at the origin of all living beings (425). But this possibility shouldn't worry us, Darwin seems to imply, because it does not invalidate the hypothesis of natural selection, based on the definition of a number of properties, not of living beings but rather of life as such. Today, it is difficult to fully appreciate the originality of this conclusion. This was the first time (or the second, if one includes Schelling) that someone managed to abstract life from its Form in such a radical way as to be able to renounce the idea of a life prototype—whether teleological, morphological, or etiological.

3.4.2. Darwin always denied having deified nature, developing old and new arguments. At some point, for example, he compares natural selection to Newton's gravitational force (63). This was an old trick. On one point, however, he was right. The god he inserts into nature,

Life, does not resemble the idols of yore; it resembles Schelling's *Grund,* an *unconscious and impersonal* ground that exacts an infinite debt of life from the living, ceaselessly instigating them to perfect themselves. In this sense, Darwinian Life is not a personal god; it is a metaphysical entity, which needs to be queried on this level, the level of a vision, the level of a "view of life," that shatters old metaphysical beliefs and replaces them with a new *metaphysics,* whose basic assumption is as simple as that: life isn't Form, it is Force.

3.4.2.1. One of the most curious aspects of Darwin's theory is the tragic exasperation of naturalistic description, which is not a matter of style but a conceptual issue. For the "struggle for life" to be more than a fantasy, it is necessary for natural selection to make *vital* choices, not just useful ones. Going down this road, however, we face significant theoretical consequences, as one can verify by checking an animal's front and rear. Let us take a giraffe. As regards the elongated neck, no doubt about it, one can consider it a *vital* characteristic that made a difference in the struggle for life and in the process of selection. But what about the tail? Certainly the giraffe's tail is useful, it has a function, it is used to chase off flies that bother the animal. And yet, it seems difficult to assign a *vital* function to the tail, for the flies cannot threaten the animal's life. This objection is not trivial. Darwin is well aware of these and other difficulties. So much so that the sixth chapter of *The Origin* is devoted precisely to their careful reexamination. But again, how does he address the problem of the tail? In his masterpiece, he gives three answers to that problem. The first does no more than restrict the theory's hermeneutical potency. Perhaps, Darwin concedes, "not everything" is explained by natural selection. But the fact is that "not everything" could turn out to be too much, in the sense that it could seriously undermine the supposition that natural selection is a *law* of nature. Thus, with the second answer, which prolongs the first, Darwin thinks things over and resorts to an *ad hoc* hypothesis. Each organ that is apparently superfluous and devoid of a vital function could have had one in the past. The giraffe's tail, which today is used to swat away flies, at one time probably did possess a vital function, perhaps it was used to chase off poisonous and deadly insects (157, 178 ff., the deadly insect example is mine). As everyone can easily see, we suddenly enter the realm of the most unchecked fantasies. No one will ever know if the giraffe's ancient ancestors needed to struggle against cyanide flies. No one will ever be able to corroborate or refute a similar hypothesis, which could be termed a "metaphysical speculation." It is therefore not surprising that Darwin rethinks the issue yet again and provides a third answer, which pushes his conception to extremes, allowing him to *see* something strange. In a passage marked by profound uncertainties, Darwin *sees* that chasing away flies is something that is authentically vital for giraffes. "The tail of the giraffe looks like an artificially constructed fly-flapper; and it seems at first incredible that this could have been adapted for its present purpose by successive slight modifications, each better and better fitted, for so trifling an object as to drive away flies; yet we should pause before being too positive even in this case, for we know that the distribution and existence of cattle and other animals in South America absolutely depend on their power of resisting the attacks of insects: so that individuals which could by any means defend themselves from these small enemies would be able to range into new pastures and thus gain a great advantage. It is not that the larger quadrupeds are actually destroyed (except in some rare cases) by flies, but they are incessantly harassed and their strength reduced, so that they are more subject to disease, or not so well enabled in a coming dearth to search for food, or to escape from beasts of prey" (157). This exemplary excerpt of Darwin's prose demonstrates two things. First, it shows that when Darwin speaks of "utility," as in the pages that follow the passage we just quoted (159–64), he is actually thinking of something more radical: he has *vital* in mind. The tail is useful; it helps the giraffe to chase off flies that bother it, but natural selection is insensitive to utility. It is sensitive to vitality. Everything that does not possess a vital meaning or value becomes *trifling* from the point of view of natural selection. It becomes insignificant and negligible. Second, this passage proves that Darwin was tormented by doubts after all. This is the reason he will acknowledge, after a few years, that sometimes he had exaggerated. "I now admit, after reading the essay by Nägeli on plants, and the remarks by various authors with respect to animals, more especially those recently made by Professor Broca, that in the earlier editions of my *Origin of Species* I probably attributed too much to the action of Natural Selection or the survival of the fittest. I have altered the fifth edition of the *Origin* so as to confine my remarks to adaptive changes of structure. I had not formerly sufficiently considered the existence of many structures which appear to be, as far as we can judge, neither beneficial nor injurious; and this I believe to be one of the greatest oversights as yet detected in my work" (Darwin 2009, 152–53).

3.4.2.2. Darwin's persistence in employing the word "useful" explains why historians, philosophers, and scientists insist on interpreting his thought in terms of "utilitarianism," sometimes mentioning David Hume's name. Darwin certainly read some of the latter's essays (Richards 2003; Flanagan 2003). But what ties exist between Hume and Darwin? And can they really shed light on the "utilitarian" lexicon of *The Origin*? To begin with, let us recall the definition of "utility" given by Hume. "Utility is only a tendency to a certain end" (Hume 2010, 67). But how can this "certain end" be known? This question prompts Hume to make a crucial distinction between "reason" and "sentiment." Reason in fact can and must evaluate the means by which we can obtain the useful, but only sentiment can point toward the ultimate end, the utility of our actions—and this sentiment, in the final analysis, amounts to a sentiment of pleasure or displeasure. "It appears evident that the ultimate ends of human actions can never, in any case, be accounted for by reason, but recommend themselves entirely to the sentiments and affections of mankind, without any dependence on the intellectual faculties. Ask a man *why he uses exercise*; he will answer, *because he desires to keep his health.* If you then enquire, *why he desires health,* he will readily reply, *because sickness is painful.* If you push your enquiries farther, and desire a reason *why he hates pain,* it is impossible he can ever give any. This is an ultimate end" (71). Hume doesn't hesitate: utility is "the tendency to a certain end," an end that is sometimes prescribed to us by reason, in the case of relative ends, and in others by sentiment, in the case of ultimate or absolute ends, which coincide with what is truly useful to us. The assumption of his entire argument is that reason and sentiment are the vectors of a will that is always oriented toward specific ends. Thus, his discourse is only valid for humans, endowed with a will that is poured into ends. Now, returning to Darwin, we encounter a diametrically opposite thesis. First of all, in his theory there is no room for "a tendency to a certain end." This would imply that natural selection wants the useful, that it is capable of discerning it, and operates with a view to its achievement. Yet, this is not what Darwin has in mind. If the struggle for utility is sensitive *to a certain end,* the struggle for life is *blind and unconscious.* Natural selection is not an intelligent principle. But there is more. Hume's reasoning, in its clarity, allows us to better grasp Darwin's reasoning. Hume says: human beings desire health because sickness is painful, and one cannot proceed further in the inquiry. In brief, the *pleasure principle* is the ultimate principle of the human will. Darwin says: living beings desire health because they desire health, because they want to live, and one cannot proceed further in the inquiry. In brief, the *vitality principle* is the ultimate principle of natural life. This entails that, hand in hand with Darwin, we end up in a region where the pleasure principle no longer has the last word, in a region that is far *beyond the pleasure principle.* Again, let us consider the fly-swatter tail. If natural selection operated on the basis of the pleasure principle, all dilemmas would vanish. The fly-swatter tail would appear unproblematically as a result of natural selection, because the flies are a nuisance, perhaps only a minimal nuisance, but still a nuisance, to animals who, over time, have outfitted themselves with this bizarre natural artifice. The tail would have been molded by selection with a view to a certain utility, *to a certain end,* dictated by the pleasure principle. Darwin, however, cannot accept such a utilitarian hypothesis, because it would covertly reintroduce final causes into biology. The category of the useful, or of "utility," would revive an old kind of finalism, namely, the teleology of forms-of-life aiming *to a certain end.* For this reason, in spite of appearance, the useful does not underlie his reasoning. And the same can be said of the pleasure principle, understood as the ultimate criterion for the evaluation of human actions and vital processes. Beyond utilitarianism, that is, we find ourselves beyond the pleasure principle, a step that, once taken, is not without consequences. The first is that health now becomes a value in itself, detached from the pleasure principle, from the value of well-being. Health is desirable, of course, but not because it is pleasurable, and illness, instead, painful. Health is simply prescribed by the vitality principle. Living beings desire health because they desire health, because they want to survive, regardless of the pleasure or pain they may feel. The second consequence is that health now cannot be qualified or evaluated on the basis of any previous criteria, which would inevitably take us back to the pleasure principle, i.e., the utility principle. Health rather turns into an empty, abstract, and indecipherable value, which comes down to the mere survival of the living being, to the mere will to live. And as soon as one attempts to say something about it, one can get confused very easily. Hence the recurring vacillation, not only in Darwin but within the entire horizon of the *salus vitae,* between the vital and the "useful."

3.4.2.2.1. It is difficult to avoid the impression that the most penetrating reader of Darwin was his most atypical interpreter, a metaphysician who was up to his times, Sigmund Freud. Whoever knows his biography knows of his unyielding admiration for the English naturalist. Whoever knows his works knows of his passion for biological theories. His essays are indeed filled with notions borrowed from Ernst Haeckel (phylogeny and ontogeny), who interacted with Darwin and popularized his views, from Wilhelm Ostwald (free and bound energy), a physicist-chemist who later worked on biological issues, and several others. But everyone knows, above all, that Freud's most revolutionary essay addresses some thorny biological problems. The essay is entitled *Jenseits des Lustprinzips* (*Beyond the Pleasure Principle*). What does Freud argue therein? As we shall see in a moment, he seeks to prove that, while keeping Darwin's theoretical framework intact, one can substitute the idea of a struggle for life with that of a struggle for extinction. In other words, Freud senses that life abstracted away from all forms-of-life and understood as a force that ceaselessly transforms organisms, creating them and destroying them one after the other, does not only manifest a vitality principle but also hides a mortality principle. This is the bridge that leads from Darwinian biology to Freudian thanatology. Let us cross it.

3.4.2.2.2. In the sixth chapter of *Beyond the Pleasure Principle* Freud introduces the notion of *Todestrieb,* "death instinct" or "death drive," and argues against Weismann, an early proponent of Darwinism. Weismann had coined the concept of "germ-plasm," that he considered to be the receptacle of variations (today we would rather speak of *genetic* variations). In his view, natural selection intervenes on the germ-plasm, which has to be neatly distinguished from the "soma" (today we would rather speak of *genotype* and *phenotype,* respectively). Weismann thought that there was some communication between these two levels of the living being, but only in one direction, from the germ-plasm to the soma, never in the other, from soma to germ-plasm (this was actually the first formulation of what is still considered the central dogma of molecular biology, according to which the phenotype cannot transmit information to the genotype). In this fashion Weismann had once and for all swept away the soft inheritance hypothesis (of acquired characteristics), which he believed to be at odds with Darwin's original insight. With Weismann "neo-Darwinism" was born, that is, a cleansed version of the theory of natural selection, which in Darwin's texts was still marred by some imperfections. Freud starts to discuss Weismann's arguments by recalling the latter's distinction between germ-plasm and soma. He stresses that for Weismann the soma is mortal, while the germ-plasm is immortal, being transmitted from generation to generation, in the germinal cells of higher organisms. But, Freud suddenly asks himself, if this is the case, why do we die? If the germinal plasm is immortal, why would the organic soma not be immortal as well? For Weismann the answer would have been obvious. In brief, the germ-plasm is like a tornado, which devastates the soma in its blind and everlasting variation and reconfiguration of living forms, of organic and somatic profiles. Freud however is not satisfied with Weismann's answer; on the contrary he maintains that, if we die, it is because a tendency to die, an unknown and imperceptible death drive, is inherent in the germ-plasm. His hypothesis, which he himself describes as a "myth," as a "fantastic" more than a "scientific" explanation (Freud 1961, 51), could be recapitulated as follows: if the germ-plasm decrees the death of all organic soma, if the vital force decrees the death of all vital forms, made transitory by the constant variation and selection of new organic configurations, then the germ-plasm itself is dominated by a deathly impulse, by virtue of which it tends, not to survive and transmit itself in the soma but to extinguish itself, gradually canceling all evolutionary achievements, all its fleeting organic projections. Said in a slightly different way, Freud argues that there is no way to decide whether in the course of natural selection a struggle for survival or a struggle for extinction is at work. "Our expectation that biology would flatly contradict the recognition of death-instincts has not been fulfilled" (43).

3.4.2.2.3. This is not exactly the way Freud develops his argument. It is not so much in evolutionary terms as in terms of organic architecture that he illustrates life's enigmatic tendency to disappear, to vanish. Higher, more complex, organisms tend to return to where they came from. This means that they tend to subdivide into their primitive, germinal, elements, which in turn explains why death appears at the higher levels of organic construction. Yet, Freud doesn't stop here. He also states that we cannot ascribe such a tendency to return—which drives organized beings to die and splinter into their component cells—only to superior organisms (Weismann's thesis). One needs to ascribe it to cells and elementary organisms as well; one needs to attribute it not only to the soma but also to the germ-plasm (Freud's thesis). And it is this point that he engages in a duel with Darwin. There exists, Freud says, only one "living substance" (*lebende Substanz*), which separates into two levels, the germ-plasm and the soma. The drive, the tendency, the instinct to return does not belong solely to the soma; it also belongs to the plasm. In this sense, it is not a characteristic of living beings, now elementary, now complex, but rather a characteristic of life itself, of life as such. "*The aim of all life is death*" (32). From this it logically follows that everything that is alive and organic tends to return, not to a lower level of organicity but to a much more radical stage of inorganicness, to a homeostatic and undifferentiated state of complete material inertia. If we admit that *"an instinct is an urge inherent in organic life to restore an earlier state of things"* (30), and if we admit that *"inanimate things existed before living ones"* (32), the conclusion is inevitable. The goal of all that is living is death, the return to a nonvital or prevital condition. Hence two queries: First, if life itself is characterized by the tendency to return to a condition of inorganic homeostasis, if such is the secret of life, if living beings in and of themselves will to die, then how did life originate on earth? What countervailing force conquered the immanent tendency of living beings to disappear? Second, if all living beings are dominated by a death instinct, by an untamed, innate tendency to vanish, how can we explain their evolution? For, at this point, we can no longer rely on a vitality principle that pulsates in all living beings and explains their infinite diversification as well as their gradual selection.

3.4.2.2.4. As far as the first query is concerned, Freud candidly concedes that there is no answer in a perspective like his, which makes of life an irresistible race toward death. "The attributes of life were at some point evoked in inanimate matter by the action of a force of whose nature we can form no conception" (32). As far as the second query is concerned, on the contrary, he feels he can give an answer, although he doesn't make it explicit, in order not to expose himself to sharp criticism. The question of evolution is addressed at the end of the fifth chapter (just before the exchange with Weismann), when Freud excludes that evolutionary processes follow the straight line of a gradual improvement of living beings. "Biology teaches us that higher development in one respect is very frequently balanced or outweighed by involution in another" (35). From this one can infer that Freud was not a Lamarckian. Having said this, it would be preposterous to see him as a carefree Darwinist. A little earlier, in fact, he writes that "we have no longer to reckon with organism's puzzling determination (so hard to fit into any context) to maintain its own existence in the face of every obstacle" (33). As a matter of fact, Freud avers, we should look for a different explanation of the evolutionary process, thus leaving aside the principle of vitality and the idea of an unconscious self-affirmation of life, which underlie Darwin's theory. But what could an alternative explanation of evolution be? Freud hints at it when he recounts, with visionary words, the beginnings of life on earth. "It was still an easy matter at that time for a living substance to die; the course of its life was probably only a brief one, whose direction was determined by the chemical structure of the young life. For a long time, perhaps, living substance was thus being constantly created afresh and easily dying, till decisive external influences altered in such a way as to oblige the still surviving substance to diverge ever more widely from its original course of life and to make ever more complicated *détours* before reaching its aim of death" (32–33). In this passage, what from a Darwinian point of view would be described in terms of environmental pressure—a passive pressure that is not able to actively orient evolutionary processes but bends now here now there under the sway of selective fluctuations—acts as an obstacle not to vitality and the living's self-affirmation, but to mortality and the living's "effort to annihilate itself." Because of some mysterious "external influences," Freud argues, the living substance had to increasingly complicate its paths in order to reach the destination toward which it tends and precipitates, death. It is the same hypothesis as Darwin's, which Freud now turns inside out. According to Darwin, if life evolves without end, it is because it wants to live and reproduce as much as possible, *but it never knows beforehand how to accomplish this task,* for it is not endowed with any intelligence or foresight, and therefore blindly searches for ever-new solutions. According to Freud, on the other hand, if life evolves without end, it is because it wants to die and extinguish itself, *but it never knows beforehand how to accomplish this task,* for it is not endowed with any intelligence or foresight, and therefore blindly searches for ever-new solutions. "Hence arises the paradoxical situation that the living organism struggles most energetically against events (dangers, in fact) that might help it attain its life's aim rapidly—by a kind of short-circuit. Such behavior is, however, precisely what characterizes purely instinctual as contrasted with intelligent efforts" (33). Having cast aside all intelligent principles, it no longer matters whether living beings, transforming themselves, evolving, are driven by a blind yearning for life or a blind yearning for death. The only thing that matters is the blindness of their yearning (which Freud, like Fichte, calls *Trieb*, "drive," "impulse," and not *Instinkt*). As a result, we cannot ascertain whether it is one or the other yearning that beats in the hearts of all living beings. We cannot know, because it is life itself that doesn't know. Life, abstracted away from living beings and their "utility," becomes so diaphanous that it can exchange roles with death, without this implying the disappearance of natural selection, which only needs to choose between two opposing opinions and can also affirm death instead of life. Living beings are born and die, being transformed from the one into the other, because living beings are breaches, passages, through which life searches for itself, without really knowing whether it is life or death. This is the extreme, albeit paradoxical, outcome of the principle of ignorance, which carries us far beyond the pleasure principle, an outcome that, one hundred years after Goethe's worries, still frightens everybody. "We should consequently feel relieved if the whole structure of our argument turned out to be mistaken" (38).

3.4.2.3. The tragic rigidity of Darwin's speculative scheme becomes evident also in another of its crucial junctions. What are the origins of society? Is it possible to account for them in biological terms? It would be possible if we could infer *altruism* from the hypothesis of natural selection. For, in order to live in society, sometimes one has to sacrifice one's own interests in favor of others. At first glance, however, natural selection seems to exclude altruistic behaviors, just as it seems to exclude stable forms of social cohabitation. The struggle for survival, which is unleashed with particular intensity not so much at the inter-specific level, as on the intra-specific level, implies living beings' radical egotism: living beings struggle individually among themselves in order to leave the largest amount of progeny. If this is how things stand, individuals are in competition with one another and cannot feel any real sympathy for each other. Accordingly, it seems impossible that they form a stable society. Darwin soon notices this snag. In *The Descent of Man,* he seeks to solve the problem. Let us suppose, Darwin says, that single individuals are not the only target of natural selection, but that intra-specific groups, such as "races," "populations," "tribes," "nations," are as well. If this were the case, more cohesive, organized, and efficient social groups would overcome less cohesive, organized, and efficient ones in the eternal struggle for survival, and in this manner we could explain the origin and development of social behaviors, of altruistic behaviors, while still holding on to the hypothesis of a basic, hidden egotism of all life-forms. But let us now abandon Darwin's line of reasoning. Let us complicate the problem and suppose that, due to stochastic variation, inside a social group of altruists there appears a free-rider, an egotist who tends to take advantage of others. Obviously the egotist will be able to act unopposed, without resentment on the part of the altruists, thus leading to the birth of other egotists, who will in their turn take advantage of the situation and proliferate abundantly, until the entire group will be invaded by egotists who will put an end to society. This is not a vain leap of the imagination. This is the kind of objection that is still raised today against the argument of "reciprocal altruism," or reciprocal advantage, that should explain the survival of societies from a Darwinian point of view (Rosenberg 2003). The objection, moreover, is perfectly consistent with the presuppositions of Darwin's theory. And what it proves, in the final analysis, is that the hypothesis of natural selection does not tolerate the existence of two distinct targets for selection, individuals and groups: one needs to choose. Since individuals are inescapable in order to explain the tree-like speciation processes starting from individual variations and the intra-specific struggle of all against all, "group selection" appears to be, of the two, the one to cast aside.

3.4.2.3.0. In spite of appearances, Darwin seems to be aware that this is the only possible conclusion that can be drawn from his theory. Thus, in the same essay, where he famously discusses the descent of human beings from primates, he starts to double-cross the reader. On the one side, he praises the moral virtues of the more civilized "nations," waving the flag of a noble altruism, and quoting Kant (Darwin 2009a, 70, 86). On the other, he repeatedly stigmatizes the degenerative effects of an altruism that actually distorts the processes of natural selection that should always reward the stronger, able-bodied, individuals, and eliminate the weak. "We civilized men, on the other hand, do our utmost to check the process of elimination; we build asylums for the imbecile, the maimed, and the sick; we institute poor-laws; and our medical men exert their utmost skill to save the life of every one to the last moment. There is reason to believe that vaccination has preserved thousands, who from a weak constitution would formerly have succumbed to small-pox. Thus the weak members of civilized societies propagate their kind. No one who has attended to the breeding of domestic animals will doubt that this must be highly injurious to the race of man. It is surprising how soon a want of care, or care wrongly directed, leads to the degeneration of a domestic race; but excepting in the case of man himself, hardly anyone is so ignorant as to allow his worst animals to breed" (168). Although he had praised it elsewhere, altruism here becomes again a *trifling* alternative for Darwin. If "civilized races" have revealed themselves as being more powerful than "savage races," and will remain so in the future, it is because under the table they allowed, and continue to allow, nature to get her way. It is because they supported, and continue to support, life's antisocial laws, leaving the big apes and Australia's aborigines behind. These are groups whose destiny is already marked by the iron-clad trigger mechanism of natural selection: they will be exterminated. "At some future period, not very distant as measured by centuries, the civilized races of man will almost certainly exterminate and replace throughout the world the savage races. At the same time the anthropomorphous apes [. . .] will no doubt be exterminated. The break will then be rendered wider, for it will intervene between man in a more civilized state, as we may hope, than the Caucasian, and some ape as low as a baboon, instead of as at present between the negro or Australian and the gorilla" (201). The sun of natural selection thus rises to the point of illuminating our future, the future of the human race that will finally manage to get rid of its inferior and embarrassing links in the evolutionary chain. In this perspective, as Darwin openly argues, not only overseas "tribes" but also the subhumans who still contaminate civilized societies are condemned to disappear. "In regard to the moral qualities, some elimination of the worst dispositions is always in progress even in the most civilized nations. Malefactors are executed, or imprisoned for long periods, so that they cannot freely transmit their bad qualities. Melancholic and insane persons are confined or commit suicide. Violent and quarrelsome men often come to a bloody end" (172).

3.4.2.3.1. Needless to say, all this left the problem of sociability and altruistic behavior (which can be found not only among humans but also among animals) unsolved. So that, after Darwin, many others had to confront the question, alternating conjectures and refutations, in a continuous see-saw of arguments. From Vero C. Wynne-Edwards to John Maynard Smith, from Edward O. Wilson to Elliott Sober, for years biologists and philosophers have discussed "group selection," just as for years they have pondered the pros and cons of a Darwinian meta-ethics, that should enable us to explain moral behaviors in terms of natural selection (Kitcher 2003). In all the cases, the problem remains more or less the same: how can we devise models of an *evolutionarily stable strategy* capable of resisting the *free-rider* objection, and suitable for deducing the *statics* of social cohabitation from a Darwinian *dynamics*? At some point, the answer seemed to be hidden in so-called game theory. This is the idea behind one of the most controversial essays in biology of the second half of the twentieth century, Richard Dawkins's *The Selfish Gene.*

3.4.2.3.2. The idea advanced by Dawkins is seductive, not so much because of its mathematical corollaries but because of its ability to cut a Gordian knot: the target of natural selection would not be either individuals or groups but rather the bricks with which these "survival machines" are built, that is, genes. Yet, what is a gene? The answer is not obvious (Pichot 1999). The gene is a segment of DNA of variable length, whose borders can only be drawn on the basis of the functionality we ascribe to it. In this sense, Dawkins observes, the gene is a "differential" unit; it is a segment of DNA that one isolates from the rest of the filament by matching the presence of specific phenotypical characteristics to the presence of specific genotypical ones. So, if the color of the eyes is blue or brown, we can go and search for differential units of DNA that correspond to those differential properties in the organism. It is quite important to stress that the gene has a differential identity, and not a material one, since the same segments of DNA may contain several genes, superimposed, crossed over one another. The principle that where one gene ends another starts doesn't obtain. The gene's identity is differential and connected to a specific functionality. Having said this, there also exists a nonspecific, generic, functionality of all genes; there exists a universal gene function, which is that of replicating itself. The gene, as Dawkins strongly emphasizes, is a "replicator." And, given this premise, why should we exclude the possibility that, in addition to the gene for eye color and the gene for homosexuality, there exists one for altruism? There would be nothing to object to such a conjecture, nor would the previous objections against "group selection" be valid, because, in this case, what counts are neither individuals nor groups, but genes and nothing else. Altruism genes could be discovered to be perfectly "selfish genes," if altruism were to be discovered advantageous for their replication through the organisms that represent their "survival machines" (the machines by means of which they propagate and move around the world). In brief, if natural selection's target are the genes and if (taking our cue from game theory) it is proved that cooperative behaviors are much more suited to genetic replication than contrary behaviors, then the deed is done. Natural selection functions on two levels, biological and sociological. We are "selfish genes" who exploit altruism with the purpose of replicating ourselves. And what remains to be understood, if anything, is the reason the gene is so selfish as to want to replicate, whatever it takes.

3.4.2.3.3. This is difficult to understand unless one examines Dawkins's assumptions better. His goal is ambitious: finding a theoretical framework in which to insert natural selection so as to make it finally compatible with the emergence of social behaviors. In order to succeed, however, Dawkins is forced to introduce categories that do not belong to Darwin's original argument and, from time to time, deviate from its tracks. His essay begins with the description of a primal scene, the birth of life, and with the enunciation of a universal law, the law of stability. "Darwin's 'survival of the fittest' is really a special case of a more general law of *survival of the stable*." That which is stable tends to exist, that which is not tends to vanish: this is what Dawkins seems to be saying. And he is thinking first of all of the general stability of matter, of "stable patterns of atoms," like "rocks, galaxies, ocean waves" (Dawkins 2006, 12). The primal scene follows, the birth of life. He writes that at a certain point in the primeval soup molecules with the unprecedented property of self-replication were born. For this to be possible, they had to absorb smaller molecules, then put them into sequence so as to make them into copies of themselves, and finally let them go away. In what measure all this is consistent with the law of the stability of matter, described previously, is not clear. So much so that Dawkins notices this and admits that at some point "suddenly a new kind of 'stability' came into the world" (16). We could perhaps define the first stability, that which pertains to the world of matter devoid of life, as the inertial stability of forms, while the latter, that which pertains to the vital world of genes, might be termed the identitarian stability of the Self. To what extent does this new stability of the selfish gene cast some light on the phenomena of life? Up to a point: sexuality. If in fact cellular division (*mitosis*) allows a set of genes to reproduce themselves faithfully, with no losses, from generation to generation, the genetic reshuffling imposed by sexual reproduction only allows one to save half of the genes in the following generation (*meiosis*). This violates the gene's selfishness, which always ought to tend to replicate itself integrally, in its totality. Dawkins, therefore, is forced to confess: "What is the good of sex? That is an extremely difficult question for the evolutionist to answer" (43), since, broadly speaking, it is difficult, if not impossible, to reconcile the *dynamics of vital aseity* (*salus vitae*) with the *statics of vital identity* (*conservatio vitae*). It is as though we were trying to reconcile Darwin and Buffon, the "transmutation" of species and the "matrix" of organisms.

3.4.2.3.4. Let's return to the gene, the "selfish replicator." If the gene replicates, and its identity is purely functional, we can apportion this identity on two levels, according to Dawkins. At a deeper and more general level the gene's function is to replicate itself, while on a more superficial and specific level the gene's function is that of better meeting the demands placed by the environment (here the gene's environment includes other genes, i.e., the *gene pool* in which a gene happens to find itself). If there is compatibility between a gene's particular function and the particular demands placed by the environment, the gene keeps on replicating and spreads around the world. But at this point, adopting Dawkins's line of reasoning, we can opt for two alternative descriptions of the same phenomena, one local, the other global. The first tells us that, perhaps due to genetic variation, replicators that are functional to the demands placed by one particular environment surface, and that they then replicate. The second says that in any case natural selection operates in such a manner as to favor those replicators that demonstrate they are more functional than the others in terms of self-replication. The facts accounted for by the two descriptions are identical, but the first is a local observation, which simply records the superficial functionality of a replicator with respect to one *particular* environment; whereas the second is a global explanation, which exalts the replicator's deep functionality insofar as reproduction in *general* is concerned. In the first case one is talking about *specific* and *concrete* types of interlocking between living forms and environmental forms; in the second one is talking about a *generic* and *abstract* tendency to the replication of all genes, which struggle for survival. The disparity between the two options is obvious: only the second one claims to explain the facts, on the basis of the universal principle of natural selection, whereas the first just records the occurrence of some events, without taking sides. How rigorous is the second explanation, which is, ultimately, the only explanation at stake? In order to answer, let us draw a more stringent comparison between the two options. If the first is an observation that just registers the *fact,* each time local and superficial, of the replication of a particular gene, in circumstances that are themselves particular, the second is an interpretation that instead tends to dramatically accentuate the global and profound *value* of the gene's survival in general, in circumstances that are themselves generalized. If the first is thus an observation that relativizes the event of replication and includes it in the *modus operandi* of a certain gene with respect to a certain environment, the second is an interpretation that absolutizes the *modus essendi* of survival and releases it from the superficial functions of response to the environment, which are seen as deriving from a single deep function: replication, reproduction. These are two alternative ways of presenting things, but a true Darwinist, an "ultra-Darwinist" like Dawkins, cannot be satisfied with the first, since there is no mention of natural selection, which *on principle* implies a struggle for progeny, a polemic between adversarial queries for survival, between rival claims for life, with a decisive choice for one or the other. In passing from the first to the second option, as should be clear, we pass from a merely observational register to a highly speculative one, which heavily overdetermines the empirical *fact* of replication, that in the first option had been recorded as an event *chronologically* subsequent to an effective environmental adaptation, and sublimates it into the theoretical *value* of survival, useful in the second option to *logically* explain all successful environmental adaptation. From this point of view, as someone has keenly remarked, "everything flows from competition among genes, or at least among organisms, to leave more copies of genetic information to the next generation. [. . .] Ecosystems, according to Dawkins, at least, are ultimately to be understood as products of the competition within and among populations for reproductive success" (Eldredge 1995, 5–6). This is how, adopting Dawkins's perspective, a fact is changed into a value: by converting survival, replication, which at first glance is only an effect (among many) of the processes of readjustment of the various ecosystems, into the (final) cause of these same processes and the added value of living beings, which on this basis can then be labeled as "good" or "bad," "better" or "worse," depending on whether or not they follow their vocation, thus following or not following the instruction that life gives all living beings. The genes, Dawkins says, cannot but "give the survival machine a single overall policy instruction: do whatever you think best to keep us alive" (Dawkins 2006, 60). In the light of all this, no wonder that on the opposite front of contemporary biology the reply has been sharp and ironic: "I, to the contrary, see reproduction as a physiological luxury rather than an imperative that is necessary for that fox to go on living" (Eldredge 1995, 40).

3.4.2.3.5. It is nonetheless true, Dawkins could insist, that the gene replicates and that its replication is the sole criterion for defining life in general, life as such. But the question remains: why do we need to define at all costs, and to concentrate on life as such? Let us consider again the principle according to which *all that replicates is alive* and let us imagine we can design a software, named "Dawkins," whose main function will be that of replicating *ad infinitum* the name "Dawkins" on a computer screen. Could we say that "Dawkins" is alive? As a matter of fact, this is what Dawkins has in mind. "The only kind of entity that has to exist in order for life to arise, anywhere in the universe, is the immortal replicator" (Dawkins 2006, 266). So, if we shipped off "Dawkins" to wander the universe, could we state that life exists somewhere in deep space? Technically we could state this if and only if we endowed "Dawkins" with a supplementary force, with a force we already encountered, the *force of the Self.* This is the force that would allow "Dawkins" to struggle for its reproduction against other replicators, and it is by virtue of this force that we could affirm that "Dawkins" is really alive, namely, that it is *selfish.* It is thanks to this force that, "as time goes by, the world becomes filled with the most powerful and ingenious replicators" (265), that is, with good genes, characterized by "a high survival value" (193). Unfortunately, the force of the Self is invisible, albeit propulsive. We cannot qualify it, we can only quantify it. For instance, we could quantify it if a virus named "Gould" suddenly entered our computer, attempting to replicate itself at the expense of "Dawkins." At that point, whoever replicates the most could claim to be "the most powerful and ingenious." Yet, what would the reason for this claim be? The reason would always be the same. The reason is that one *abstracts* one aspect, replication, from the entire process, and that one makes of that aspect the *purposiveness* (without a purpose) of the entire process. What do we gain from this? In reality, we add nothing new to our description of vital processes. All we have done is to define them as, precisely, *vital*. We have made a mathematical measurement of a set of phenomena, a mathematical measurement that is meaningless *per se* (we can in fact measure any set of phenomena in a lot of different ways), and we have glued a word on it, "life," "replication," "survival," seeing the word as an explanation of the thing in itself. It is difficult not to speak of idealism in this case. Difficult not to see that the selfish gene is, willy-nilly, an idealist gene. In the end, doesn't the same *force of the Self* pulsate in Dawkins's gene and in Schelling's *Geist*?

3.4.2.3.6. Thus, at the very core of this view of *vital* phenomena we find what has provocatively been termed the *Allmacht,* the omnipotence of natural selection (Gould 2002), which is currently being subjected to tough scrutiny by biologists. On the microevolutionary level, the discovery of a number of genetic mutations that remain neutral and insensitive to natural selection has led to theorizing a "non-Darwinian evolution," characterized by processes of a stochastic nature (Motoo Kimura). On the macroevolutionary level, results from paleobiology have persuaded most people that selective flows are not able to explain those sudden mass extinctions that certainly took place in the past (David Raup). On both levels, as on many others, natural selection sees its ambition, that of embodying a theory of biological Everything, being severely limited (Gayon 2003). And since Darwin's hypothesis is the only one that allows us to talk about "life" in the abstract, at least from a scientific point of view, it is precisely this concept, "life," that seems to be called into question today, even in the field of biology, even by the science of "life." Hence the query: why do we still remain tied to this little word, "life"; why do we continue to use it in scientific contexts, Dawkins being a prime example, and more generally in the sphere of everyday and public discourse? The answer to the first half of the question could be technical in nature: "life" perhaps functions as an "epistemological indicator" *à la* Foucault, namely, as a metaconcept that unifies observations, notions, theories that would otherwise look scattered, messy, and completely disconnected. But the answer to the second half of the question could be another one, and could cast a clearer light on the first half of the question as well: "life" is probably the only name we manage to give ourselves, to our Self, to our humanity, to our freedom. If this were the case, this would confirm once again that we, the moderns, are still on the way to "autonomy," and that we are still crossing the bridge that takes us from Kant to Darwin: to be ourselves means that we are free (Kant), but to be free means that we are simply alive (Darwin). As Dawkins put it in his own words, our genes are "free and independent agents" (Dawkins 2006, 62).

3.4.2.4. The problem of the unit, or of the "target," of natural selection is such a frequent object of debate because it is among those that are insoluble within a Darwinian perspective. The unit of selection ought in fact to be redefined as the aseity of selection. And this is a difference not to be overlooked: *the One is not the Self.* Who (or what) is natural selection's target? Not the One, a form-of-life, but rather the Self, a force-of-life, which wills to live, which demands to reproduce as much as possible and which continues to grow stronger in the course of evolution. For this reason there is no (unequivocal) unit of selection: because, upon closer inspection, there is no longer a unit that has a stable ontological status from a Darwinian point of view. Living beings, organisms, genes, social groups are reduced to transitory forms, moved and swept away by an invisible force. The world of forms-of-life is reduced to a foam, brought to the surface by the deep and savage wave of the force-of-life. It is on this that natural selection applies its blind filtering. Darwinian selection promptly provides an answer, *life or death,* to a question that has the same content, *life or death.* But this content cannot be qualified, it can only be quantified. It does not express a form, it only expresses a force. Natural selection does not give its consent to a *form*-of-life, for this would imply that it can see it, that it can have some, however minimal, comprehension of it. If selection gives its consent, it gives it to a *force*-of-life, not to this or that unity, quality, or individuality of living beings. So, the problem of the unit, or of the target, of natural selection is destined to remain unsolved. Darwinian selection does not operate on the unit(y) of life, it operates on the empty and diaphanous aseity of life. And it does not even operate on it. It rather names it.

3.4.3. Two consequences, two inexorable laws, follow from this split between unity and aseity, between living beings and life. We have already encountered them in the context of Schelling's *Naturphilosophie.* They now come back to the fore in the frame of Darwinian speculations. The first is *the law of life's imperfection.* If "life" surfaces as a demand for life, "life" will never be fulfilled, satisfied with itself. "The life of the animal and the plant bears the same universal character of incompleteness as the life of man. This is directly attributed to the circumstance that nature—organic as well as inorganic—is in a perennial state of evolution, change and transformation" (Haeckel 1934, 217–18). The second is *the law of human beings' imperfection.* If "human beings" are living beings, then "human beings" are defaced, in their natural humanity, by a congenital defectiveness. "It is a vulgar prejudice that Human Beings are the most perfect of living organisms and they represent in all their body parts and all their functions the acme achieved by biological evolution" (Morselli 1904, 375). Both laws had been anticipated by Darwin. "No country can be named in which all the native inhabitants are now so perfectly adapted to each other and to the physical conditions under which they live, that none of them could be still better adapted and improved" (Darwin 2009b, 64–65). *Der Mensch ist ein ewiger Bruchstück*; human beings are eternal fragments, would have been Schelling's gloss.

3.4.3.1. The semantics of modern life thus seems to stem from a single metaphysical root and to split into two branches, representing two traditions of "thought" (*Denken*): the tradition of continental thought, marked by an idea of "life" that Schelling was the first to announce; the tradition of Anglo-Saxon thought, marked by an idea of "life" of which Darwin was an equally strong, and perhaps stronger, proponent. That the lymph in both traditions is provided by the same concept of "life" is something that becomes all the more evident when one looks more closely into the "thought" of two philosophers who, apparently, are divided by an ocean, Daniel C. Dennett and Georges Canguilhem. Both, despite appearances, talk the same language, the language of late modernity, a language in which "life" becomes a synonym of freedom, but there are no longer any human beings who can advocate it.

3.4.3.2. Dennett's thought is exemplary from many points of view. His basic idea is that philosophical concepts should be *naturalized,* that is, dissolved, wherever possible, into scientific concepts, or else abrogated. The science Dennett is thinking of, however, is not just any science; it is a specific science, born from "Darwin's dangerous idea" (Dennett 1996). It is the science of natural selection. The task of contemporary philosophy is therefore to naturalize the concepts over which the philosophers of the past have racked their brains, such as the concepts of consciousness or freedom. What is consciousness? What is freedom? Today we should answer these questions with the instruments of Darwin's theory, perhaps revisited by Dawkins.

3.4.3.2.1. Consciousness, for instance, is not what Descartes believed it was, a *res cogitans* fluctuating some place, which mysteriously interacts with a *res extensa,* with our body, with our cerebral or physiological states. Actually there is no such thing as a *res cogitans.* The only thing that exists is the *res extensa.* And, in order to understand what consciousness is, all we need is to grasp how that *rex extensa* that our organism is functions. In a nutshell, Dennett's thesis is the following: in our brain there exist various "agencies," each of which proposes behavior patterns to our body, which compete with one another; of the many proposals only one is chosen each time by the organism, and this is the one that temporarily gets control; but then the process continues in our brain, and power changes hands, from one "agency" to the next. In light of these dynamics, what is our Self, our consciousness? It is a higher-level process, Dennett maintains, a process of *narration,* thanks to which the work of the many cerebral "agencies," distributed over time, is made coherent, to the extent possible, so as to make my behaviors justifiable and communicable to other people. The Self, that which makes *myself,* that which allows me to refer to *myself*, is therefore a cognitive representation that is capable of integrating my acts and making them decipherable, interpretable, for others. In the final analysis, my Self is of a social nature. Yet, we could ask ourselves, why are we, humans, social beings? Because, Dennett replies, that is what our evolutionary history decreed; because, in the past, sociable behaviors and communication skills proved to be crucial factors in the survival of human beings. "A proper human self is the largely unwitting creation of an interpersonal design process in which we encourage small children to become communicators [. . .] A *person* has to be able to keep in contact with past and anticipated intentions and one of the main roles of the brain's user-illusion of itself, which I call the self as a center of narrative gravity, is to provide *me* with a with a means of interfacing with myself at other times. [. . .] The acts and events you can tell us about, and the reasons for them, are yours because you made them—and because they made you. What you are is that agent whose life you can tell about. You can tell us and you can tell yourself. The process of self-description begins in earliest childhood, and includes a good deal of fantasy from the outset" (Dennett 2003, 273, 253–54). Here, to put it in a few words, we find ourselves in front of our "personal" Self, that we might also call our *cognitive Self*. On this level, Dennett argues, the Self is an illusion, a narrative, a fiction, whose purpose is simply that of favoring social work, group work, which is obviously aimed at increasing our survival capabilities, given that Dennett is a convinced Darwinist. But things don't end here, since another Self exists in addition to the cognitive one. Dennett calls it the "self-as-ultimate-beneficiary." We could rename it the *vital Self*. This further Self is extremely important in Dennett's view, for it is the target of natural selection, like Dawkins's selfish gene. For Dennett, the personal Self and the vital Self do not coincide for many reasons. First of all, because the vital Self, being the target of natural selection, must be placed at the level of genes and even of social groups (following a slightly revised version of Dawkins's argument). Yet, the rationale for such a sharp distinction between the personal Self and the vital Self is much more cogent than that and goes far beyond their outward appearances. A vital Self is, in general, anything that makes itself a candidate for natural selection. Each time we raise the question *Cui bono?* we are reasoning in terms of natural selection, and we should therefore find a vital Self that can embody its target. From this it follows that there is an ontological divide between the vital Self and the personal Self. The vital Self, in fact, cannot be seen as an illusory fiction, like the cognitive Self, since it is the foundation, the everlasting protagonist of natural selection. It is its *subjectum*. This is the reason Dennett's theory can ultimately be described as a theory of the *double Self*. "There has to be something in the role of the self—something that defines the answer to the *Cui bono?* [. . .] A self-as-ultimate-beneficiary [i.e., a vital Self] can in principle be indefinitely distributed. I can care for others or for a larger social structure, for instance. There is nothing that restricts me to a *me* as contrasted to an *us*. [. . .] It is better to think of the human capacity to rethink one's *summum bonum* as the possibility of extending the domain of the self. I can still take my task to be looking out for Number One while including under Number One not just my own living body, but my family, the Chicago Bulls, Oxfam . . . you name it" (180). Dennett's reasoning is faultless, and it confronts us with the existence of two Selves, a real Self and an illusory Self, the vital Self and the personal Self, a Self that is the rock-like foundation of natural selection and a Self, the human *person*, that is only its belated fruit. But then, where has *our* personal freedom ended up? Does it exist or is it an illusion? And if it exists, is it an attribute of the vital or of the personal Self?

3.4.3.2.2. Needless to say, if freedom is not an illusion, it can only hide in the vital Self; if it were an attribute of the personal Self, it would be as illusory and fictitious as the latter. Freedom, if it exists, must therefore coincide with the force-of-life that agitates all forms-of-life, with the simple demand for life that each living being expresses, *in its-Self*, so as to be exposed to natural selection's verdict. In this sense, Dennett remarks, freedom has existed in the world since life has existed. The similarity with the idea advanced by Schelling is striking. "Life is the autonomy in the phenomenon; it is the scheme of freedom, insofar as it reveals itself in nature" (Schelling 1980b, 222). Two centuries later, Dennett writes: "The freedom of the bird to fly wherever it wants is definitely a kind of freedom, a distinct improvement on the freedom of the jellyfish to float wherever it floats, but a poor cousin of our human freedom. Compare birdsong to human language. Both are magnificent products of natural selection, and neither is miraculous, but human language revolutionizes life, opening up the biological world in dimensions utterly inaccessible to birds. Human freedom, in part a product of the revolution begat of language and culture, is about as different from bird freedom as language is different from birdsong. But in order to understand the richer phenomenon, one must first understand its more modest components and predecessors. What we must do to understand human freedom is to follow Darwin's 'strange inversion of reasoning' and go back to a time at the beginning of life when there was no freedom, no intelligence, no choice, but only proto-freedom, proto-choice, proto-intelligence" (Dennett 2003, 143). Therefore, Dennett argues, "there are all manner of different kinds and grades of freedom" (94) depending on the evolutionary rank, superior or inferior, of the different forms-of-life, and there exists a precise criterion to evaluate the rise, the "improvement" of forms-of-life in the course of evolution. The criterion on whose basis living beings "improve" is the gradual increase of freedom within nature. An increase that in turn coincides with a rise in the possibilities of survival. What is the freedom Dennett is speaking about in fact? It is the capability to avoid danger, which all life-forms dispose of. This capability can be stronger or weaker. In any case, the organism's freedom coincides with its capability, with the "power" it has to reduce the margins of "inevitability" of natural processes that hang over it and can cause its death. Hence freedom in nature increases the more the options living beings have to avoid danger(s), shaping their behavior one way or the other. The more options living beings have at their disposal, the more possibilities to survive they have, and the more freedom they have. That is, the more "autonomous" from natural dangers they are (142–43). "We need to take the etymology 'inevitable' seriously. It means *unavoidable.* Curiously, its negation is not used, but we can easily enough coin the term, and note that some things are *evitable* by some agents, and some things, in contrast, are not evitable by those agents" (56). Now, one of the most interesting implications of this argument (developed in order to make determinism compatible with freedom) is that danger, for Dennett, can be "evitable" only within the purview of an agent that incurs *that specific* danger. What represents danger for a jellyfish will not necessarily do so for a shark or for a horse, nor for human beings. The "evitability" of danger always implies a perspective and *subjective* framing of the danger on the part of the living. So much so that Dennett is forced to conclude that the horizon of all living beings, from the simplest to the most complex, is "*subjectively* open" (91). Thus, according to Dennett, life is freedom, life is subjectivity, and this life is that of a vital Self that does not amount to any personal Self, to any cognitive Self, i.e., to consciousness. This life is, to all intents and purposes, an unconscious aseity that pulsates in the depths of all living beings. But then, once more, where has *our* freedom ended up? What difference can there ever be between the generic freedom of all living beings and the specific freedom of human beings?

3.4.3.2.3. Among all beings, humans are the freest, Dennett proclaims. This means that from his point of view the human is the form-of-life endowed with the greatest number of "subjective" options. They are so numerous, the degree of human "evitability" is such, the spectrum of choices that human beings have at their disposal is so vast, that one can no longer speak of a "human nature." It is rather as if humans had been modeled by natural selection so as to continuously change their nature, to mold it, adapting it to the countless environmental dangers they must face. Dennett makes the point clear: "Our natures aren't fixed because we have evolved to be entities *designed* to change their natures in response to interactions with the rest of the world" (93). In this case, too, the similarities with Schelling are astounding. Just as for the German idealist the human species embodied an empty identity, a peak of absolute indeterminacy that nature reached in the course of its evolution, suddenly opening itself to History, so for the American evolutionist humans find themselves at the crossroads of *nature* and *nurture,* to the extent that humans lose a nature of their own, a "human nature," and become a peak of indeterminacy, of infinite "evitability" that natural evolution reaches, opening itself to History. In human beings, in fact, natural "evitability" becomes so wide and profound that *natural* (or "genetic") evolution, in order to continue on its path, is forced to change level and transmute itself into an *ideal* (or "memetic") evolution, into a historical, social, cultural evolution of ideas (of "memes"), in which a new type of "evitability," of freedom, tends to impose itself: a cultural "evitability" or freedom, with selection processes that will no longer be natural, but cultural themselves. So, from natural selection we move on to cultural selection (from genic selection we move on to memic selection), always under the same battle's banner, always in the name of survival, first of natural forms, then of cultural forms (first of genes, then of memes) that both vociferously demand to reproduce and spread. Only the "best" will prevail, only those best adapted to *live,* because we are still moving within the framework of Darwin's view of life. Cultural selection remains, that is, a natural selection. Both engage in the same type of *vital* selection. That said, who will be victorious in this cruel battle for life engaged by cultural forms? Logically, as Dennett explains, it will be that social, historical, cultural group that manages to bring forth the greatest "evitability," the greatest flexibility, thus being capable to avoid the greatest dangers. This is the reason the culture that manifests the flexibility or "evitability" of human beings to such an extent as to exclude the very existence of a "human nature" will be the one that wins the battle for survival. Inevitably, this will be the culture of "autonomy," the culture of absolute freedom even from natural bonds, in its most up-to-date and powerful variant: the *Darwinian culture.* The extermination of inferior savage *races,* imagined by Darwin, is thus replaced by the extermination of inferior savage *cultures.* Although Dennett doesn't use the word extermination, the logic remains exactly the same. "In the next century it will be our memes, both tonic and toxic, that will wreak havoc on the unprepared world. *Our* capacity to tolerate the toxic excesses of freedom cannot be assumed in others, or simply exported as one more commodity. The practically unlimited educability of any human being gives us hope of success, but designing and implementing the cultural inoculation necessary to fend off disaster, while respecting the rights of those in need of inoculation, will be an urgent task of great complexity, requiring not just better social science, but also sensitivity, imagination, and courage. The field of public health expanded to include cultural health will be the greatest challenge of this century" (304).

3.4.3.3. Canguilhem's view is different, and in certain respects the opposite. He is well aware of the "selfish" outcomes of the modern abstraction of life; it is clear to him that this abstraction can lead to questionable practices both in scientific and political contexts; he knows full well that some of the twentieth century's disasters conceal the cruel revenge of life against the living; and he senses that life has been gradually changed into an idol in the course of modernity. So much so that his attempt will be precisely that of absolving life from its modern interpretation, to make it something more tolerable, more *livable*. An interesting attempt, certainly not because of its success, but because of its failure, which assumes the value of a warning.

3.4.3.3.1. One could synthesize Canguilhem's basic assumption by quoting the answer that Claude Bernard, a contemporary of Darwin, loved to give when he was asked: what is life? "I don't know, I never met it," was his reply. This could be used as an epigraph for Canguilhem's works. What is life in the abstract, life in general? One cannot tell, he would have explained, because statistical or typological "norms" of life cannot be established. A "normality" of life that can apply to this or that class of living beings, if not even to all living beings in general, cannot be predictively fixed. This precept, which lies at the heart of Canghuilhem's reflections, is meant to push us to question the modern notion of life, in its most generic and abstract sense, which regularly tends to cast the needs, the "utility," and even the dignity of the living to one side. Canguilhem, in contrast, insists on the concrete and binding singularity of each living being. That said, are his arguments sufficient to make us overcome the modern semantics of life? Difficult to believe so, if one examines the texts attentively. The idea, for example, that life does not express, in its explosive dynamics, a "normality" that can be attained once and for all and that can be determined *a priori,* according to statistical or typological (or further) criteria, is an idea that already runs through the modern tradition and frames its reading of vital processes, as we have seen before. If we substitute the word "normality" with the word "perfection," one does not need much to realize that Canguilhem, in reality, falls right back into a view of life that unfailingly restates the same axiom: life itself, life as such, contests the existence of any "normality," of any "perfection," in living beings. We called it the *principle of perfectibility.* Nothing substantially changes when it comes to another tenet of Canguilhem's philosophy, that is, the rediscovery of the "subject in biology"—an idea that Kurt Goldstein, frequently quoted by Canguilhem, had already advanced in his time, as had Viktor von Weizsäcker, an author he quotes less but is an important reference nonetheless. In this case, too, although one guesses the benign intentions and the desire to trace life as such back to the "subjective" perspective of individual living beings, thereby overcoming life's cold and cruel anonymity, one can easily realize that the infusion of subjectivity into life proposed by Canguilhem is already one of the cornerstones of the modern tradition. It would be difficult to deny, sharpening one's gaze, that life's vitality, in its modern epiphany, comes to coincide with the most intimate subjectivation of living beings, nailed to a Self that is their secret strength and their life itself. We called it the *principle of unconscious aseity.* All these subtle affinities between Canguilhem and his predecessors, which seem to undermine the originality and the critical incisiveness of his statements, become literally embarrassing when one gets into the third principle, the third invisible keystone of the modern parsing of life. We called it the *principle of variability.* It is mainly around this principle that Canguilhem's reflections revolve, and it is here that we run into serious problems concerning his theory, which finally proves much less innovative and emancipatory than he believed it to be. It is not possible to talk about life in the abstract, Canguilhem maintains, because it is not possible to find out general norms, whether statistical or typological, of life itself. Accordingly, no life abstracted away from individual living beings really exists, because each living being establishes the parameters of its normality *on its own,* constantly varying its own vital norms and measures. In that regard, one might possibly speak of a kind of *individualistic vitalism.* "That is to say, in dealing with biological norms, one must always refer to the individual because this individual, as Goldstein says, can find himself 'equal to the tasks resulting from the environment suited to him,' but in organic conditions which, in any other individual, would be inadequate for these tasks. [. . .] When it comes to a supra-individual norm, it is impossible to determine the 'sick being' (*Kranksein*) as to content. But this is perfectly possible for an individual norm." (Canguilhem 1991, 181). Yet, the question then arises as to how one can establish, each time, *one* vital norm in the singular. What is this "individual norm" of life? How can we define it? According to Canguilhem, one cannot define it *in the abstract,* once and for all, because such an "individual norm" is dictated by a "normative capacity," by the *normative force* of those living beings that are capable of building up a dynamic and flexible relationship with the environment, thus continually updating and diversifying their norms of adaptation. Hence the definition of "health" and "sickness," the first in terms of normative capacity, the second in terms of normative incapacity; the first in terms of force, the second in terms of weakness of the individual when confronted with environmental challenges. The sick individual is the one who does not have sufficient normative force to inventively widen his margins of adaptation to the changing circumstances and conditions of life. This is the individual who scleroticizes his conduct and reactions in front of the ever-changing stimuli of the *milieu* in which he is cast. "The sick living being is normalized in well-defined conditions of existence and has lost its normative capacity, the capacity to establish other norms in other conditions. [. . .] The patient is sick because he can admit of only one norm. To use an expression which has already been very useful to us, the sick man is not abnormal because of the absence of a norm, but because of his incapacity to be normative. [. . .] What characterizes health is the possibility of transcending the norm, which defines the momentary normal, the possibility of tolerating infractions of the habitual norm and instituting new norms in new situations. [. . .] To be in good health means being able to fall sick and recover: it is a biological luxury" (183, 186, 196–97, 198–99). To sum up: for Canguilhem life is a normative force that stands behind the organism and makes it more or less flexible, adaptable, *variable* when confronted by the challenges it must face. That said, the question is: to what degree does this view of life differ from the modern abstraction of life? To what degree does life *qua* normative force differ from life *qua* "formative force" of "organized beings"? To what degree can we domesticate modern life, and make it more livable, if we continue to see it as an invisible Force that stands behind living creatures?

3.4.3.3.2. To respond appropriately, we need to examine how Canguilhem treats the problem of health. According to him, health is not a *static* condition, nor a specific psycho-physiological state that tends to "maintain itself." On the contrary, health is a *dynamic* expression of life; health has to be seen as a power of "expansion." Health is that "organic vitality" of living beings that corresponds to their "capacity [. . .] to establish a new order" and "desire to dominate the environment" (201). In this sense, just as life in the abstract does not exist, health in the abstract does not exist either. If to live means to continuously redefine one's relationship with the environment, varying and adjusting organic norms and psycho-physiological equilibria with respect to environmental situations that also vary continuously, health understood as an immobile condition, as an average or typical state of this or that organism, does not exist. There is no such thing as a natural health that can be recorded and relaunched as a fixed parameter for the evaluation of this or that living organism. If in nature there existed something like health in the abstract, organisms would be able to reach internal equilibria, distance themselves from them and then perhaps return to them, oscillating around a condition of *static normality,* which is instead rejected by Canguilhem, who is the theorist of the *dynamic normativity* of living beings. In brief, health does not exist in nature. Living beings are never "in health." Or better, if we say that they are "in health," it is because they are going through a process of never-ending "healing," of pressing and hectic readjustment of their organic and psycho-physiological parameters. This is why, if health still exists somewhere, we cannot find it among the *facts*, according to Canguilhem, but among *values*. The fact of health does not exist, in his view, but the value of health still does. Health is a value that dictates the perpetual "healing" of living beings, their dynamic and propulsive remodulation of their vital capabilities. Here, the tone of Canguilhem's voice becomes almost Nietzschean: "Health is a feeling of assurance in life to which no limit is fixed. *Valere*, from which value derives, means to be in good health in Latin. Health is a way of tackling existence as one feels that one is not only possessor or bearer but also, if necessary, creator of value, establisher of vital norms" (201). At this point we are in a position to spell out, without entering into too many details, the basic principles of Canguilhem's philosophy of life: (a) Life is parsed in the singular, it is always the life of *one* single organic individual, of *one* single psycho-biological entity endowed with a "personality" of its own (184). (b) Life does not express a *static normality*; it expresses a *dynamic normativity*, which coincides with the capacity of the living being to create unprecedented norms of vitality. (c) Health does not exist in nature, if by health we understand a predetermined normality (established on the basis of statistical or typological criteria) of living beings; health is rather a sort of unstable and highly inventive reconfiguration of the organism that is capable of "healing" (Canguilhem 2012, 53–66). (d) Consequently, health is not a natural fact but rather a "value," the value of life, the ultimate and absolute value that life expresses as a normative force, as a healing force, as a force that creates continuously renewed and always variable, diversifiable, values (43–52). From this it follows that a science of "health" cannot exist in Canguilhem's perspective, in exactly the same sense that a "science" of value cannot exist in Kant's perspective (44–45). The two arguments mirror one another, and the rationale behind them is identical: it is impossible to seize the value of an individual's will and life by means of predictive and scientific knowledge, because the willing and living individuals have value *in themselves* only to the extent that they can express their will and life *by themselves*, in a spontaneous, variable, and perfectly individual manner. Just as for Kant the will of the moral individual manifests the *absolute* value of the autonomous will, that is a will capable of willing *by itself,* unpredictably, that which it wills, likewise for Canguilhem the life of the organic individual manifests the *absolute* value of the autonomous life, that is a life capable of living *by itself,* unpredictably, that which it lives. In neither case is a "science" of this value possible, for this value amounts to the absolute autonomy of individual will and life, which stands behind individuals. This value is that which makes them individuals; it is the *a priori* that makes them what they truly are, willing and living in their individuality, even before they can open their eyes and realize, as soon as they do so, that they can't even see what they are.

3.4.3.3.3. The outcome of such speculations, admittedly metaphysical more than scientific, is not that innocuous, despite Canguilhem's noble intentions. If life is a normative force that permeates all living beings—now more, now less—making them capable of flexibly adapting—now more, now less—to the novelties they encounter, and if the degree of normative force—now greater, now lesser—is what makes all living beings live, that which makes them more or less autonomous from environmental conditions, *individuating* and *freeing* them from external bonds, from heteronomous constraints, then one can conclude: first, living beings have to forcefully struggle, in the course of their existence, so as to individuate and free themselves from all that impedes them in their progress; second, by struggling every day, living beings never feel they are sufficiently individual and free, since they remain engaged in the struggle; third, this drive toward self-individuation and self-liberation, toward the autonomous creation of a profile of one's own, places them in a state of inevitable tension against themselves, against the inertia of their pasts, from which they strenuously tend to separate; fourth, the motor for this tension against themselves turns out to be the value, no longer the fact, of health; fifth, health thus shines as an absolute value, as a cogent imperative. In light of this, it is not surprising that Canguilhem defines health as the "unconditioned," mimicking Kant's definition of freedom. The two terms in fact, health and freedom, become synonyms at this point, intoning the imperative that one must categorically obey, *even going against oneself.* This is the categorical, "unconditioned," imperative of autonomy. The one who wills, the one who lives, must obey it, but in order to do so one has to say goodbye to oneself. "Let us call this health free, unconditioned, unaccountable" (49). The old Kant had already made this point, muttering his last meditations. But in addition to Kant, two hundred years later, the shadow that returns is above all Schelling's, who had announced the "ever higher tension" of living beings against themselves with quite another degree of vigor, picturing the pains of a natural hell, which is ours (Schelling 2006, 66). The dry prose, not at all inclined to flights of rhetoric, does not prevent Canguilhem from placing all living beings' adventure in a scenario that is equally dramatic, where they are condemned to chase an unachievable health, condemned each day to heal from one knows not what malignant threat. "The so-called healthy man thus *is not* healthy. [. . .] The menace of disease is one of the components of health" (Canguilhem 1991, 287). This is the not very reassuring outcome that we end up with, contesting modern life, attempting to disentangle ourselves from its invincible abstraction. Although Canguilhem sought to oppose its implacable demands, the final result is still the one implied by the *metaphysical logic* of modern life. The result is "us," sick human beings, who are defective in both health and freedom, due not to any contingency whatsoever, but to an ironclad metaphysical necessity. The same necessity that deprives us, *a priori,* of "humanity." Throughout modernity, the motto cannot but echo again and again "There is no human nature" (Sartre 2007, 22). Thus, our battle cry keeps on resounding: *Salus vitae.* And for human beings like us, for humans who can no longer call themselves "human," because they can no longer give this word any meaning, the only appropriate label is the one provided by Canguilhem himself, the aristocratic rebel, with irony and bitterness. These are two words with which, three centuries earlier, another famous philosopher had addressed his fellow citizens. *Ultimi barbarorum.* "Spinoza apparently made a mistake. Believing that the barbarians he was denouncing were the last. But he knew Latin. He meant: the latest, the most recent" (Canguilhem 1990, 32).

3 US

ON THE USE AND ABUSE OF LIFE FOR HISTORY

"The stronger must dominate and not mate with the weaker, which would signify the sacrifice of its own higher nature. Only the born weakling can look upon this principle as cruel, and if he does so it is merely because he is of a feebler nature and a narrower mind; for if such a law did not direct the process of evolution then the higher development of organic life would not be conceivable at all.

This urge for the maintenance of the unmixed breed, which is a phenomenon that prevails throughout the whole of the natural world, results not only in the sharply defined outward distinction between one species and another but also in the internal similarity of characteristic qualities which are peculiar to each breed or species. The fox remains always a fox, the goose remains a goose, and the tiger will retain the characteristics of a tiger. The only difference that can exist within the species must be in the various degrees of structural strength and active power, in the intelligence, efficiency, endurance, etc., with which the individual specimens are endowed. It would be impossible to find a fox which has a kindly and protective disposition towards geese, just as no cat exists which has a friendly disposition towards mice.

That is why the struggle between the various species does not arise from a feeling of mutual antipathy but rather from hunger and love. In both cases Nature looks on calmly and is even pleased with what happens. The struggle for the daily livelihood leaves behind in the ruck everything that is weak or diseased or wavering; while the fight of the male to possess the female gives to the strongest the right, or at least the possibility, to propagate its kind. And this struggle is a means of

> *furthering the health and powers of resistance in the species. Thus it is one of the causes underlying the process of development towards a higher quality of being.*
>
> *If the case were different the progressive process would cease, and even retrogression might set in."* (Hitler 1939, 222–23)

Faced with these words the normal reaction is to turn one's eyes elsewhere. And in doing so today's biologists, perhaps disturbed by the use of the expression "natural selection" in the *Protocol of the Wannsee Conference* (the reunion in the winter of 1942 during which the higher echelons of the Nazi hierarchy decided to proceed with the extermination of Europe's Jews), usually follow three lines of reasoning: they say that Charles Darwin's words don't coincide with those of Adolph Hitler; they say scientists are not responsible for the use(s) others make of their theories; they reiterate "natural selection's" innocence accusing the National Socialists of having badly distorted Darwin's theorems (Gould 1996, 309–24). As far as the first two arguments are concerned, there is no doubt that Hitler himself would have subscribed to them, even boasting of it. Hitler's political ideology goes well beyond Darwin's statements as a naturalist. But can we say for this reason that he *distorts* them? Or shall we say that he *saturates* them? And supposing this to be the case, supposing that Nazi ideology saturated and resolved Darwinian equations by assigning a previously fixed value to some of their variables, can we take this saturation to be a sheer distortion? Are we really sure that this is the right way to understand, with a clear mind, what occurred seventy years ago? Are we really certain that "to saturate" is a synonym of "to distort"? And why, moreover, was there such a tenacity on the part of the German establishment of the 1910s, 1920s, and 1930s in distorting Darwinian theorems about life specifically?

If we resist, even only for reasons of caution, the temptation to immediately discard these queries, we can better approach something that for historians is now a matter of record, one that unfortunately has not yet stimulated the attention of philosophers as much as it ought to. *From Darwin to Hitler* is the title of two historical essays, published recently, which confront the thorny issue of Hitlerism's derivation from Darwinism openly (Pichot 2009; Weikart 2004). But what does "derivation" mean here? The question is no longer historical; it is purely philosophical. Historians, for their part, limit themselves to registering the contiguity and sometimes the obvious kinship between documents scattered in the dust of facts, but the *logic* on whose basis these documents can, or must, be placed one after the other is something that it is philosophy's duty to inquire about. "Darwinian terminology and rhetoric pervade Hitler's writings and speeches, and no one to my knowledge has ever questioned the common assertion by scholars that Hitler was a social darwinist. It is too obvious to deny" (Weikart 2004, 7–8). What remains to be understood, therefore, is to what extent Hitler had distorted Darwin's and Darwinists' thought, completely misrepresenting it, and to what extent, instead, he held it steady in the background in order to articulate a theory and practice of politics that Darwin, certainly, could never have imagined, but which still has some ties, however tenuous, with his hypotheses. That Darwin was an intellectual hero for Nazi ideologists, and before that for German militarists, like general Lothar von Trotha, who, at the beginning of the twentieth century, planned and realized the century's first genocide, the extermination of the Hereros in southwestern Africa, is something that is now considered established fact by historians. The colonialists rallying cry at the time was "racial war," and for them, as for Houston Stewart Chamberlain, the teachings to be followed were those of a teacher who was not Nietzsche or Wagner or Gobineau. "My teacher is none other than Charles Darwin" (Chamberlain 1912, 14). Through the mediation of people who have gradually been removed from our memory, like Ludwig Gumplowicz, Gustav Ratzenhofer, Ludwig Woltmann, Georges Vacher de Lapouge, Otto Ammon (Lukács 1981; Traverso 2003), or like Friedrich Ratzel (the inventor of *Lebensraum*, "vital space," and of the *Kampf um Raum*, the "struggle for space"), Klaus Wagner and Sebald Rudolf Steinmetz (two leaders of Darwinian militarism), and also thanks to the cover provided by well-known and respectable scientists like Oscar Peschel, Friedrich Hellwald, Moritz Wagner, Max von Gruber, Felix von Luschan, Fritz Lenz, Oscar Schmidt, Ludwig Büchner, Ernst Haeckel (Gasman 1971) and many more, Darwinism became an integral part of the German cultural-political language at the beginning of the twentieth century, preparing a terrain on which the extremists of the National Socialist party would soon march.

Having said that, once more, how much of Darwin actually remained in such a context? Although all dikes were breached, something remained, namely the abstraction of life itself, of which Darwin had become the patron saint. There remained life raised to the dignity of ultimate value of modernity, of that "us" in which we still recognize ourselves. This obviously does not mean that Darwinism and Hitlerism are the same thing. Rather, it means that without Darwinism Hitlerism would not have been possible. Without Darwin Hitler and his comrades would have had to invent different words, different thoughts, different ideas. This is a historical truth, which is in one sense banal, but in another certainly requires serious and thorough consideration. The Darwinian natural world has been—it is difficult to dispute this—a necessary condition for the irruption of the Hitlerian political world. Not a sufficient but a necessary condition. And why is this important? *Precisely because Darwinism and Hitlerism are not identical.* The world of Darwinian life was already there; it had become our world already before Hitler's advent, and, above all, it is still our world, the world of autonomous life. Let us therefore align these three statements: (a) the Darwinian natural world has been a condition of possibility of the Hitlerian political world; (b) this condition should be understood as necessary, not sufficient; (c) this condition is still with us today. At this point it becomes fairly urgent to protect oneself with regard to the links that the logic of autonomous life seems to impose on all of "us" in conceiving ourselves and others, in conceiving what is human and its manner of being in the world, if we don't want to risk stumbling on our own feet once more, in the dark.

> *Under appropriate direction the Jews are to be utilized for work in the East in an expedient manner in the course of the final solution. In large (labor) columns, with the sexes separated, Jews capable of work will be moved into these areas as they build roads, during which a large pro-*

> *portion will no doubt drop out through natural reduction. The remnant that eventually remains will require suitable treatment; because it will without doubt represent the most [physically] resistant part, it consists of a natural selection that could, on its release, become the germ-cell of a new Jewish revival. (Witness the experience of history.) (Protocol of the Wannsee Conference, January 20, 1942)*

Focusing on this document it is possible to frame the problem also from a different point of view. Today many people ask themselves questions on the topic of biopolitics, on the meaning and origins of this theoretical and practical paradigm of politics that for some completes, for others overturns, and for yet others replaces the classical grammar of sovereignty. In philosophical terms it becomes a question of outlining the framework within which a theory and practice of biopolitics become possible, if not inevitable. This frame can be deemed to be more or less ample and inclusive. Let us take Giorgio Agamben's and Roberto Esposito's approaches. For both of them the Nazi extermination camps are the tragic emblem of the contemporary biopolitical drift and for both, even if from partially different perspectives, it is mandatory to reconstruct their genealogy.

For Agamben the camps are to all effects and purposes the secret of Western sovereignty, a secret to be taken almost in the etymological sense of the term, for sovereignty was sooner or later destined to *secrete* the camps; it was only a matter of time. The camps, and all that is condensed there, are in fact something that was latent in our juridical system from the very beginning. The horizon in which such an interpretation moves is thus that of the entire "Western" tradition, which extends to the point of including Roman law, where the figure of the *homo sacer* is already creeping about. Agamben, as is well known, leads the category of "bare life" back to this figure, that he pairs and counterposes to the concept of "sovereign power." In this way, he tends to adopt a very wide-ranging perspective that lowers itself from on high, vertically, onto (at least) two thousand years of "Western" history. But does the West exist? And if it does, what is it? Agamben, like many others, believes it does exist and believes, following Heidegger's teachings, that the West is our remote past, our provenance. Given this premise, he can coherently recount his reading of the camps, which are deciphered as the obligatory outcome of that matrix-apparatus of the entire Western tradition that is the law. Let us call it the camps' juridical grid (Agamben 1998).

For Esposito the camps cannot be exclusively read in these terms. One needs to address the problem from a different point of view, and fashion new interpretive tools, which can be used to restrict and adjust our perspective. Very succinctly, in order to understand how we arrived at the camps, we should not raise our gaze to the level of the West's entire past, we should only look into the history of modernity, which in his opinion and that of many others begins with Hobbes. The question doesn't change; it is still a question of understanding in what manner sovereignty *secretes* biopolitics, the deadly hold of sovereign power over bare life. But the sovereignty in question is a more recent political category, whose inextricable interweaving with further categories—such as freedom and property—needs to be questioned. All these categories, which are the pillars of modern political thought, are *double face* according to Esposito, in that on their *verso* they hide a biopolitical aspect, against whose background the thanatopolitical turn of the camps becomes possible, at least tendentially, thus making that definitive superimposition of the domain of sovereign power and the domain of biological life that is the *political* stigma, not so much of the West, as of the moderns, possible in the long term. Let us call it the camps' political grid (Esposito 2008).

The undoubted merit of these philosophical interpretations of the camps lies in their attempt to make that which for a long time was pushed into the realm of the "unexaminable," of the unthinkable, thinkable. "When some say, casually, that what the Nazis did (the extermination) is of the order of the unthinkable, or the intractable, they forget something crucial: that the Nazis both thought and treated what they did with the greatest care, the greatest determination. [. . .] Nazism is both a politics and a thought" (Badiou 2007, 4). That said, the question then immediately arises as to what extent biopolitical interpretations of the camps hit the mark and really put the Nazi "thought" underlying Nazi "politics" into focus. For, in the final analysis, this "thought" took roots not only in a more or less remote past, but also in a "century," the twentieth, that opened with a very specific ontological question: "At the beginning of the twentieth century the dominant ontological question is: what is life? [. . .] What is real life? What does it mean to truly live, live a life adequate to the organic intensity of living? This question traverses the century and should be related to the question of the new man. [. . .] The thought of life questions the force of the will-to-live" (18).

To be perfectly clear: while Agamben insists on the camps' juridical grid and Esposito insists on the political one, I think it would be better to insist on the underlying metaphysical grid of the camps, the grid that has allowed all of us to use the word "life" in a certain way, not over the course of the last two thousand years, or the last four hundred years, but over the last two centuries. If in fact it is true that "bare life" now towers at the center of the juridical and political apparatuses, as highlighted by Agamben and Esposito, it is also true that "bare life" is not a self-evident concept. Life, "bare life," *bloßes Leben,* is a recent invention. Paraphrasing Foucault we could say that life, *this life,* is the print that remains in the sand once the wave has swept away the print of all human faces. And Foucault himself, moreover, with whom all theorists of biopolitics still confront themselves, was well aware of the fact that the thought of life, precisely in its Darwinian sense, remained the uncrossed threshold of his many excavations, aimed at reconstructing the genealogy of Nazi totalitarian racism on multiple levels. "Basically, evolutionism, understood in the broad sense—or in other words, not so much Darwin's theory itself as a set, a bundle, of notions (such as: the hierarchy of species that grow from a common evolutionary tree, the struggle for existence among species, the selection that eliminates the less fit)—naturally became within a few years during the nineteenth century not simply a way of transcribing a political discourse into biological terms, and not simply a way of dressing up a political discourse in scientific clothing, but a real way of thinking about the relations between colonization, the necessity for wars, criminality, the phenomena of madness and mental illness, the history of societies with their different classes, and so on. Whenever, in other words, there was a confrontation, a killing or the risk of death, the nineteenth century was quite literally obliged to think about them in the form of evolutionism" (Foucault 2003b, 256–57). But why all this? Why was it precisely Darwinism, which at the end of the nineteenth century was not even a hegemonic paradigm within the international community of biologists, that became the most stringent form of discursive articulation of a racial biopolitics? And to what degree did this form influence, and perhaps still does influence, concrete political practice, orienting it in certain directions and making it credible, acceptable to all of "us"? Why have we been "literally obliged," as Foucault underlines, to think in this manner about political struggle and confrontation? It is probably asking himself these kinds of questions that Heidegger, more than thirty years earlier, in the full delirium of war, had developed the following reflections, which should have marked his distance from Nazi ideology.

> *"[. . .]the metaphysical ground of the sciences is occasionally taken notice of as such, and admitted, and then forgotten again; at other times, however, it is mostly not thought of at all, or is rejected as a philosophical chimera.*
>
> *When certain predominant views in biology about living beings are transferred from the realm of plant and animal life to other realms of beings, for example, that of history, one can speak of biologism. This term designates the already mentioned extension—and perhaps exaggeration and transgression of boundaries—of biological thinking beyond its own realm. Insofar as we see an arbitrary misuse here, an unfounded violence of thinking, and ultimately a confusion in kinds of knowledge, we must ask what the reason for all this is.*
>
> *What goes wrong in biologism, however, is not merely the transfer and unfounded extension of concepts and propositions from the field proper to living beings to that of other beings; what goes wrong already lies in the failure to recognize the metaphysical character of the propositions concerning the field, propositions by which all biology that is genuine and restricted to its field points beyond itself. Thus biology proves that, as a science, it can never gain power over its own essence with the means at its disposal. Biologism is not so much the mere boundless degeneration of biological thinking as it is total ignorance of the fact that biological thinking itself can only be grounded and decided in*

the metaphysical realm and can never justify itself scientifically." (Heidegger 1991, 45)

| | | |

Let us now leave biologism aside and try to descend to greater depths. Behind biologism Heidegger actually perceived the disturbing shadow of "nihilism." Yet, what is "nihilism," this strange monster that he is not the only one to see as representing the devious profile of our times? It is either erasure of values or flattening of values or multiplication of values, which creates disorder in our ethical and political landscape. Depending on which nihilist option one prefers, one can either speak of values' relativism (Nietzsche), values' polytheism (Weber), or values' pluralism (Berlin), which respectively contradict the absolutism, the monotheism, and the axiological monism of ancient religions. And in the name of such nihilism, with its multiple declensions, one can even claim that our freedom is greater than that of times gone by, and that we now enjoy complete autonomy, as Kant would have said. In the name of such nihilism, that is, it seems that we are no longer nailed to the idols of premodernity. The problem, however, is that our nihilism does not express an integral absence of values, nor does it express, if one looks carefully, a relativism, polytheism, or pluralism of values. Rather, it expresses an absolute value, the value of life, the value of the will, which coincide in their *autonomy*. This is why it would perhaps be better to speak of the absolutism, monotheism, and monism of autonomy. This is the nihilist religion of our freedom. Take, for instance, the two contemporary meanings of freedom, from which we previously started, the freedom *from* and the freedom *to*, negative and positive freedom. What happens in the world of autonomy? What happens is that freedom is conceivable only in negative terms, as freedom *from*, as a freedom that voids itself of all content, letting us float in a space that doesn't impose any bonds, but does not offer any directions to our choices either. Our freedom thus becomes an absolutely blind freedom, which provokes disorientation and loss of bearings. Our freedom becomes a white flag, devoid of any sign. And this is where life intervenes: to saturate this void, not with a fullness but with another void. Life comes to offer protection and orientation to our freedom *from*, giving it the (ineffable) semblance of that strange freedom *to* that is the freedom to live. A freedom that is just as empty as the first but that appears to give some tangible content to our autonomy. This is the metaphysical circle of our freedom. "Being" then becomes a synonym of "being alive." And to be alive means to heighten, strengthen our force-of-life, burning all forms-of-life, demolishing all qualifications of life, despoiling it of all attributes that are not its pure diaphanous aseity. This is "us," alive and free in our autonomy, alive and free in a new religion, which like the religions of the past dictates the gradients of axiological intensity of the facts we confront, making some visible and others invisible, or enriching and impoverishing the epistemic value of notions and categories that all begin to rotate around the hard core of autonomous life. What follow are scattered examples, however crucial, that would, obviously, require an altogether different treatment.

a. *Racism*. It is sufficient to take Darwin literally to make racism an inevitable appendix of the metaphysics of autonomous life. It is sufficient to line up the statements that, in his opinion, describe the origin of species. First, "race" is the subspecific variety of each species. Second, there exist an ever-increasing number of races, some of whom win the battle for life, which others on the contrary lose. Third, the winning races in their turn become species that, with the passage of time, develop further races (or varieties). This is, very succinctly, what *The Origin of Species* states. And starting from here, if we are logically consistent, "race war" (the *Rassenkampf* theorized soon afterward by Gumplowicz) appears as one ingredient of natural evolution, perhaps disagreeable, yet essential. Biological racism, which even before assuming a political echo is a corollary of the hypothesis of natural selection, becomes inescapable at this point. This is not the only, but it is without a doubt one of the most consolidated and recurrent justifications of racist political ideology in modern times: Darwinism, i.e., the theory of natural selection, is racist and discriminatory, as facts have demonstrated many times and unfortunately continue to demonstrate (Herrnstein and Murray's book *The Bell Curve* is one of the most recent examples of this). That said, one may ask why so many scholars of racism, even authoritative ones, tend to glide over this original sin or tend to diminish its—especially historical—significance (Mosse 1978; Fredrickson 2002). But this is a question that would deserve a separate inquiry. There seems to be something here that we have difficulty distancing ourselves from.

b. *War.* Race war is not a war like others, like those that were known before Darwin, namely, a war that interrupts peace for a more or less long period of time. Quite the opposite, race war is a *global and permanent war* (Galli 2010), like those that are promoted today, and this because Darwinism is a polemological ontology, it is a metaphysics of power, in whose context peace literally becomes inconceivable. If there were peace, this would mean that the battle for life and the struggle for life had been interrupted. Which would mean that natural selection ceases to operate. Which would mean that life ceases to pulsate. This is why in a similar metaphysical framework the war for life cannot but become ever harsher: because it is in war that, each time, *Homo darwinianus* finds his freedom. This had already been the prerogative of *Homo hegelianus,* and it remains the prerogative of modern human beings even after the metaphysical schema of the "struggle for recognition" and the "clash of civilizations" has lost all its bite. Darwinian man does not need to appeal to a civilization to start his war; it is sufficient for him to appeal to a race. He does not even struggle for pure prestige, because he only struggles for life as such. During the twilight of all social and national ideologies, there the freedom offered by the belligerent religion of life still remains. So, if it is true that for Hegel "peace is opposed to freedom" (Momigliano 1990, 141), this is true to an even greater extent if, after Hegel, we look to Darwin.

c. *Environment.* If we make the speculative framework of natural selection ours down to its foundations, everything that we could classify as environment secretly becomes *invisible.* Natural selection is not, in fact, an environmental selection. Natural selection does not follow any preestablished direction, not even the environmental one, and the word "environment," moreover, is not a part of Darwin's technical lexicon. What is the environment for him? It is everything that surrounds the organism, its environs, which can be subdivided into three distinct spheres: the physical, geological, and climatic conditions of existence; individuals of other species; individuals of the same species. Each single organism, each form-of-life, enters into relationships with each of these environmental spheres, and each single individual, in effect, *sees* a specific environment of its own, which nevertheless remains invisible to life as such. By means of variation and selection, life gradually shapes unprecedented forms-of-life that will *see,* with time, environments that are themselves unprecedented. Thus, the environment, or a certain composition of environmental conditions, exists for a specific form-of-life. But *the* environment does not exist for the generic force-of-life that unleashes the transmutation of all living beings. The environment is rather shaped, modeled, by that force-of-life by means of ceaseless reconfiguration of the forms-of-life, which confront ever diverse and renewed environments. From the fish to the amphibian and from the reptile to the bird environments change, but life, which transforms the fish into the amphibian and the reptile into the bird via variation and selection, in and of itself *does not see* any environment. Final causes, in sum, do not exist in biology, as the most consistent Darwinists remind us; there do not exist any final forms that evolution must approximate; there do not exist any forms of this kind that might perhaps have been sculpted, prefigured, outlined by the environment. Darwinian evolution does not have nor can it have any direction, not even one dictated by the environment. Rather, the environment is the malleable surface on which, as in a negative, the passing profile of the various forms-of-life—subject to a vital evolution and not to an environmental one—come to be printed in reverse. In this sense it is as if the environment were planned, designed, architecturally devised, almost dreamed by this or that form-of-life. It is as if it were the "extended phenotype" of the living unit (Dawkins

1982). So that, in a world dominated by Darwin's naturalistic language, a powerful split between life and environment tends to be produced: it tends to be produced on the epistemic level, where the idea of a habitat in which living beings can peacefully settle loses substance, and it tends to be produced on the level of everyday language. For it appears difficult, at this point, to weigh the meaning and value of something, the environment, which ends up losing all ontological endurance under the pressure of a metaphysical framework revolving around the semantics and the exorbitant requirements of autonomous life. Instead, it appears inevitable that one will privilege horizontal struggle, the competition among living beings leading to constantly modified environmental dreams, over vertical adaptation—which is not necessarily competitive—of living beings to a preexisting habitat. If for Life the environment does not exist, if Life instead leverages all environmental profiles to continually feed the infinite alteration of the forms-of-life, from which eternally diversified environmental profiles will emerge, it is because Life no longer pulsates in an environmental space but rather in a diaphanous space, into which all natural habitats are swallowed up and precipitate. Thus, the environment morphs into a pure and abstract "vital space" (*Lebensraum*) that no longer knows fixed boundaries, either natural or political, but only unstable frontiers, corroded by an insistent, unstoppable "de-territorialization" (Deleuze and Guattari 1983). Borders become trenches.

d. *Body.* If one includes the Darwinian vision within an even deeper and more general metaphysical plot, it is easy to see further paradoxes rise to the surface. The dignity and quality of each single form-of-life, for example, tend to vanish into thin air within the metaphysical horizon of autonomous life. Dignity and quality ought to concern the different forms-of-life captured each time in their individuality and in their mutual difference, but the value of life as such, its axiological force, within this horizon, never coincides with the value of a specific form-of-life; vitality in effect does not coincide with the *utilitas* of single living beings. Furthermore, the metaphysical option of autonomous life peremptorily directs us toward a region that is well beyond the pleasure principle, so that, in a similar context, living beings' pleasure or their capacity to valorize life itself as a function of their own interest loses all meaning. Life's monotheism does not grant *axiological gravity* to everything we could label as *qualified life*. The only value, the ultimate, extreme value of our religion is reduced to the mere survival of bare life. As a result, the way is paved for the dilemmas that afflict the so-called bioethical debate (Rodotà 2006). Whether it is a question of fetuses, comatose states, or of anomalous figures like the "transable," everywhere a disorienting split between the living being's body (its form) and its life (its anonymous force), which respectively denote the unity of the living being (its identity and corporeal integrity) and its simple vital aseity, is produced. Human beings, redefined as autonomous lives, tend to lose the solidity of autonomous bodies, in a constant, inflexible process of "de-corporeization" (Duden 2002). Against this background the *zoe*, bare life, invades the field that was once occupied with full legitimacy by the *bios*, qualified life. This split between human life and body, a body that tends to lose its measure and its form under the pressure of a force that detaches itself from it, can be recorded in both the frontal counterposition between the autonomous life of the fetus and the woman's body, reduced to a physiological substrate of a formless embryo, as in the frontal counterposition between bare life and the comatose body of the dying, in which life continues to pulsate blindly as a will-to-live, a demand to live, that wants nothing to do with the *utilitas* or the *dignitas* of a being exhausted by suffering. But what is this life that tends to deface the confines of the living, to unravel its fabric, in the name of its "intrinsic value," sacred, inviolable (Dworkin 1993)? What are the metaphysical roots of this dilemma-like split between human life and body? In spite of appearances, we are no longer facing the relics of old religions here. To prove this, suffice it to say that the embarrassment of decision, in matters like these, does not fall on the priests of the ancient religions, but on the modern priests of life, that is to say, on medical doctors and biologists. In all these cases then, it is a question of insoluble dilemmas that are unleashed not by an old faith but by a new faith, by a new religion, which is the religion of autonomous life. The semantics of modern life, of a life that can be deciphered only in metaphysical and quasi-religious terms, aligns and joins our scientific and ordinary languages along a common front, allowing us to draw a sacred line between living beings and their lives, between a body (endowed with form and measure) and a life (which goes beyond its form and measure). Yet, once more, what is this life that can cruelly torture the human body? This is the query we answer every day by being content to survive, struggling to succeed at all costs, secretly, and perhaps regretfully, perceiving our residual freedom in it.

| | | |

These are only some short accounts, which help clarify that behind our so-called nihilism, rigorous, detailed, insurmountable epistemic and axiological hierarchies are hidden. In this perspective, where are human beings ending up? As strange as it may sound, within a similar metaphysical framework human beings' "humanity" disperses before even appearing on the scene. In fact within the horizon of autonomous life no specific essences exist or persist. For this reason, at least tendentially, the hypothesis of an unalterable human essence or human nature is doomed to vanish. The savage ontology of life is incompatible with the old essentialism; it is an ontology of the Self that unhinges and breaks down the old ontologies of the One. From "will" to "life," the Self glimmers like a *force* besieging the One, devoted to containing this uncontainable force within a *form*. The idea of natural "perfection" also vanishes and is replaced by the principle of the infinite "perfectibility" of the One, which is condemned to chase after itself, to search for itself, without ever adhering to the Self. Given these premises, it is no surprise that human beings end up lacking "humanity." This is by no means due to chance. It is rather due to a new kind of *natural* necessity.

That said, the schema of autonomous life allows one to determine some *gradients* on the basis of which one can, if needed, measure the greater or lesser force-of-life inherent in all forms-of-life, including the human. These gradients of intensity are those that we previously obtained from the Darwinian theory of natural evolution and correspond to the three main conceptual pillars of Darwin's approach: variation, adaptation, selection. These are, respectively, *variability, perfectibility,* and *vitality* (here opposed to utility). These are the gradients according to which it is possible to classify all the more or less strong, more or less powerful, more or less advanced forms-of-life, without resorting to final, rigid, essential forms. And such are also the *measures* of the human in the world of autonomous life. Human beings, in other words, are all the more human the more they appear variable, perfectible, vital—or "formable," as Fichte would have said. This means that human beings are the more human the more they appear to be alive. Thus, human beings' profile dissolves into the formless profile of life, which is the reign of pure *intensity,* devoid of specific inalterable qualities, denuded of essential properties.

Accordingly, the science of human beings dissolves into the science of life. "Human nature doesn't exist"—in the sense that by now only a *human life* exists. And the more this life is alive, and the less it proves inclined to become rigidified into a human form, into a *human essence,* which would endure and last over time. Man becomes by definition the "new man," humans become by definition "new humans," infinitely variable, perfectible, vital, since humans are life, and since life is revealed, even more generally and deep down, to be the unconscious and diaphanous aseity of all forms-of-life, its anonymous and cogent force. Hence the Kantian–Darwinian imperative that comes into effect in the world of autonomy, the categorical imperative of health and life's salvation: "Live!" This is an imperative that, however, carries with it its implicit, unpleasant, flip side, by virtue of which human beings (*de facto*) never live enough, and precisely for this reason (*de iure*) will need to live more and more, will need to increase their force-of-life ad infinitum. The radical evil that Kant discussed in relation to the moral person is now combined, if one follows this road to its conclusion, with its natural complement, the radical illness that Darwinian living beings are affected by. But what do we find at the end of the road? We don't find the banners of an impartial nihilism but rather those of an implacable racism, which is the heart of darkness of all gospels of life. Francis Galton made a utopia of it, with an eloquent title: *Kantsaywhere* (1910).

Let us once again put the statements we made earlier in a row. In the world of autonomy, man becomes the "new man," human beings become "new humans," or the humans that are naturally to come. Indeed, if humans are life, and they never live enough, humans must always become human in the diabolical circle of autonomy. As a consequence, the science of human beings is transformed into the science of human beings *to come*, a science that from now on will no longer be the old anthropology but will be the science of life itself, i.e., biology. Now, against this background, two different types of racism can develop, mutually communicating and easily interchangeable, which we could define as the racism of racial purity and a racism of human purity. Unless we schizophrenically embrace them both, a possibility that shouldn't be excluded, one will tend toward either the first or the second depending on whether one provides a full or an empty definition of the "new man." The full definition is the one that tends to give a specific historical designation to the identity of humans and life—for example the designation "Aryan"—thus establishing the *biological measure* of the superior and inferior races, tying humanity to the gradients of the biological intensity of life, variability-perfectibility-vitality, which one will be able to test according to methods preestablished and chosen, each time, by the doctors and biologists of the day. These are the measures and rates of humanity, the human degrees and potentials that were examined in the Nazi camps, with minute experiments, aimed at capturing the weaker or stronger pulsations of life behind the emaciated faces of the victims. In this case it was a matter of tearing the mask from human beings, to allow the disfiguring howl of life as such to burst forth. It is not an exaggeration to talk in this regard of "horrorism," as a new, hypermodern experience of the human (Cavarero 2009). The empty definition is instead that which tends to give an alternative designation to the identity of humans and life—the designation "Western," for example—by establishing the *metaphysical measure* of the superior race, without nailing humanity to the chains of blood and soil, and without referring to preestablished biological parameters of humanity. Here the abstract degrees and potentials of life are not fixed by opposing two biological races, although this possibility remains on the horizon, but are rather claimed as the discriminatory indexes of two metaphysical and completely abstract races: the race of those who are *free to live* and the race of those who are not yet free, the race of adults who've come of age and the race of minors, the race of autonomous humans and the race of the infirm. And why are these precisely two *races*? Because in autonomy's metaphysical space there exists only one pure and superior metaphysical race, the race of humans who are bare life, *bloßes Leben,* the race of humans who will never be a Species, but precisely a Race, to whom *life itself* forbids any further speciation. The human Race then becomes the banner of a metaphysical racism, which associates the abstraction of life with a new powerful abstraction. Who is "Western" man? Who are "we"? We are abstract individuals. We are those who are *free to live* as they please, to the extent that we believe we are even free to modify the biological texture of our humanity on the individual scale, unlike minors, unlike the infirm. We believe we are free to do so *singulatim* and according to individual choice precisely because we got rid of the dream to gradually generate a new human species from the race we are. In "us" the autonomy of life and the autonomy of the will come to coincide in every respect, as they become two sides of the same metaphysical abstraction. We are abstract human beings. We are diaphanous individuals, who are infinitely and constantly perfectible. Exactly for this reason we—who are, by definition, the *humans to come,* the *individuals to come*—are the sole *trustworthy and truthful* champions of the human being.

| | | |

In the light of the above, it seems legitimate to hazard some distinctions, which can be extrapolated by means of a Darwinian lexicon (not the only lexicon available, but one of the most fashionable in late modernity). The first, between two types of racism: the pre-Darwinian racism of originary human beings, and the Darwinian racism of the humans of future possibilities. The second, between two types of hypermodern racism: the biological and the metaphysical.

The third, between two types of permanent war: war between races and the war of the human Race. The fourth, finally, between two models of collective identification of the human, which are like the two extremes of one single metaphysical biology, the two poles of an identical compass, oriented now toward the one term, now toward the other: we will call them social Darwinism and political Darwinism.

a. *The Racism of Origin and the Racism of Destiny.* When one speaks of racism, one would always need to establish two premises. First of all, one would need to recall that racism exists from the time the word "race" was used in a public discursive context (historic, scientific, political) and that prior forms of discrimination cannot rigorously be labeled as so many forms of "racism": *strictu sensu* racism exists from the end of the seventeenth century, more or less from the time of François Bernier's essay, *Nouvelle division de la terre par les différentes espèces ou races qui l'habitent,* published in 1684. Secondly, one should recall that after the birth of naturalist taxonomy, with Linnaeus, the human genus was split into species and races following multiple criteria. Linnaeus, for example, believed that the human genus included a single species, the human precisely, which could be subdivided into six different races. For Cuvier as well humans were part of a single genus, which included a single species, in which one could then distinguish three distinct racial groups. Some years later, instead, with Haeckel, who purveyed Darwin to German soil, the division between humans starts to get more pronounced, and one starts to differentiate twelve species within the human genus, from which no less than thirty-six races branch out: discrimination becomes deeper (even human species exist) and simultaneously becomes much more detailed (human races multiply). All things considered, despite their diversity, these various forms of naturalistic racism can be matched with no more than two types of political racism (without considering speciesism, which is just an exacerbation of racism). The first type, first also in order of historical sequence, is what we could describe as the *racism of origin.* It is the racist discourse that develops in France, for instance, from Henri de Boulainvilliers, at the beginning of the eighteenth century, to Arthur de Gobineau, in the middle of the nineteenth century. It is the racism whose genealogy and development Foucault has reconstructed (Foucault 2003b), at least partially. For this type of racism, race is no more than the stock *origin* of a lineage or of what at the time one could also call a "nation." Race is a community of birth, from which some descend and others don't. This community cannot change its characteristics over the course of time; at most it can degenerate, mixing with other races, without in principle knowing either evolution nor transformation. In this case, the race can only become extinct, or perhaps worsen, weaken, as Gobineau states melancholically, but under no circumstances can it improve or strengthen, for a very simple reason: because no one has yet hypothesized that this might occur, because no one has yet imagined that a race can evolve to the point of giving origin to a new species. This someone is Darwin, and it is with him that a form of racism, unknown up to this point, which we could baptize the *racism of destiny,* surfaces. Within this unprecedented conceptual framework, race becomes the stock of the *descent* of a bloodline, which should be understood as a biological "variety," no longer as a historical "nation." Race is, this time also, a community of birth, seen however not as a community of fathers but rather as a community of offspring, a community that can change its characteristics over the course of time, or rather, that must do so at all costs. In such a perspective, in fact, the more "alive" a race is, the more it tends to improve, to strengthen, and this not on the basis of predetermined and specific criteria for each race, relating to the latter's origins, nature, or essence, but rather on the basis of some abstract characteristics and potentialities of life, which are those more or less openly pointed to by Darwin, previously summed up in the three postulates of variability, perfectibility, and vitality. In this fashion we slip from a racist *statics* of origins into a racist *dynamics* of destiny; we precipitate from a racism of the *past* into a racism of the *future*; the *memories of ancient humans* are replaced by the *project of new humans,* to be detailed in several ways (Galton, Büchner, Ammon, Vogt, Haeckel, Vacher de Lapouge), which give rise to extremely diverse and curious expressions of racism and which all share the idea of an unavoidable disequilibrium among races. In the Darwinian war for life a strong, superior race is inevitably destined to conquer and overwhelm another, weak and inferior one. This is the inexorable Darwinian law of life. Ergo, one race will be revealed as superior, the other as inferior, *tertium non datur.* But on what basis can one decree this superiority? On the basis of racial potentiality indices that will have to coincide, not with the unalterable properties of this or that race but rather with the abstract gradients of Darwinian life, with scalar degrees of variability, perfectibility, and vitality. Gradients that are only indexes of the greater or lesser local *intensity* of life. Gradients that are only rates or exponents of generic vital potentialities, detached from any binding relationship with the concrete properties of any specific race. Gradients that, whether interpreted one way or the other, on the basis of elastic and mutable theoretical parameters, can give rise to a number of *axiological variations,* even very heterogeneous ones, on the topic of the superior race. Hence, on the one hand, the multiple forms of biological racism (characterized by the insistence on one racial trait or other in which life presumably flows more than elsewhere, according to previously selected parameters in the various disciplinary contexts: sociology, criminology, psychiatry, and so forth) that flourish on Darwinian soil, with diverse and sometimes opposing ideological orientations, and yet all variants of an identical racism of destiny, an identical progressive racism, which did not exist at all prior to Darwin. Hence also, on the other hand, the multiple contaminations between the old and new racisms that become concrete possibilities at this point, against the background of that abstract painting that is the portrait of autonomous life. This is the case with Chamberlain, for instance, who mixes Gobineau and Darwin, racism of origin and racism of destiny, but then confesses, as soon as he can, his clear and understandable preference for the latter. The racism of origin does not allow one in fact to express that propulsive, bellicose, imperialistic drive that is characteristic of the racism of destiny, but instead allows both Chamberlain and many others to give the Darwinian new man a recognizable appearance, who otherwise would run the risk of being totally faceless—an obscure and formless profile of a future, that of Life, which is just as obscure and formless. Here, it is as if the clear, recognizable, reassuring profile of original man were to be superimposed on the pure, obscure, formless vital *intensity* of humans to come, of the winning and superior race, it now being established that the force of Life as such pulsates in it, a force that overflows everywhere and imposes a strengthening, a progress instead of a regress, a move forward instead of a return to origins, of the master race. So that, when the moment comes, it will not be a time of degeneration and decadence, as required by the old national racism; it will instead be a time of palingenesis or radical extinction, as required by the new metaphysical schema. If it is not one, it is the other, not only for the Jews but also for the Aryans, whose complete extermination Hitler will decree in the name of the ironclad laws of life, even though in the "Nazi myth" the future of natural History and human evolution had, for a brief period, taken on the sculptural outline of Aryan origins.

b. *Biological Racism and Metaphysical Racism.* Within these coordinates, racism therefore tends to diversify its expressions, always remaining with its feet firmly planted on the terrain of the metaphysical biology represented by late nineteenth-century evolutionism. At this point we are dealing with a biological racism more than a national racism, even though one can encounter mixed formulas that occasionally dress biological racism in the musty clothes of national racism. And we are dealing above all with a multiform racism, which draws the demarcation line between strong and weak, superior and inferior, wherever it pleases, given the degree of elasticity of the underlying doctrine's framework, whose skeleton can promote any form of political claim, any demand for oppression. Consequently there exist many variants of the racism of destiny, before and after Hitler. It is not even a given that we have experimented with all of them. The most important variant, which is also the most predictable, is the colonialist and imperialist variant, which exalts the superiority of civilized races ("us") and deplores the inferiority of the savage races ("the others"). In a passage we already cited, Darwin states something that was nothing more than a commonplace for his contemporaries, one destined to last a long time. He talks about a future time, not very far away if measured in terms of centuries, in which the civilized races will "exterminate" the savage races. At that time, he adds, the gap between humans and their similars, in other words apes, will widen. Expressed in a slightly different way, there is an evolutionary hierarchy, established by natural History, in which humans are placed higher than apes, but some human races are placed higher and others lower, closer to gorillas (Darwin 2009a, 201). As for these inferior races, Darwin remarks, their fate is already decided. They will be exterminated. This is why we could baptize this first form of racism of destiny, both colonialist and imperialist, a *racism of extermination.* Now let us suppose we draw the same demarcation line between the strong and the weak, superior and inferior, no longer in correspondence with the territorial borders that separate the civilized from the savage but rather within our societies. The result will be that the population will be divided into social groups, endowed with unequal biological attributes. On the one side we will find the more evolved, therefore superior, group, on the other a less evolved, therefore inferior, group. Not only that: the weak, less evolved group will also inevitably bear the traces of our evolutionary past, and will, in this sense, be plagued by "primitivism," a thesis that, once more following in Darwin's wake, will soon be proposed by some early Darwinists like Walter Bagehot, Ludwig Gumplowicz, Cesare Lombroso, and Georges Vacher de Lapouge (Hawkins 1997). In this case, however, don't think of extermination, because this would jeopardize social stability. A domestic extermination would moreover be inconvenient for economic reasons, not to speak of possible moral scruples. The solution is therefore a different one. The solution is the supremacy of one group over the other, from now on conceived of as the supremacy of a more "evolved" race over another that is less so. In sum the solution is the naturalization of social disparities and of social conflict, which goes hand in hand with an integral biologization and medicalization of the human condition. The disparity is biological before being social, and it is social only because in and of itself it is biological, so that the supremacy of some over others, within the same society, appears to be a natural fact—and as such unmodifiable. Here, it is still a question of biological racism, because it is in the biological data that the destiny of social supremacy or subordination of the single individual is written. Yet, racism is no longer supported by the simple evidence of cultural difference, as in the case of colonial racism. Rather, we might speak of a programmatic racism. It is in fact an entire research program on the edge between biology and the human sciences (psychiatry, sociology, criminology . . .) that is launched, followed, and promoted, against the background of a biological racism that tends to be the more and more indistinguishable from a metaphysical racism. It is a very elaborate and diversified research program, whose insinuations, wide diffusion, and hammering propaganda are well known in the entire "Western" world, starting in the last years of the nineteenth century. All that which is part of it, even today, goes to make up the landscape of a second type of racism of destiny, one that we could term a *racism of supremacy*. Now, let us finally suppose we draw the same abstract demarcation line between strong and weak, superior and inferior, no longer between civilized and savage races, between a dominant and a subjected group, but between man and woman. Even this can be done within the context of a metaphysical and eminently programmatic racism of Darwinian ascendancy. And how can one therefore be surprised if a world-renowned biologist, about thirty years ago, stated an idea like the following: "In hunter-gatherer societies men hunt and women stay at home. This strong bias persists in *most* agricultural and industrial societies and, on that ground alone, appears to have a genetic origin. [. . .] My own guess is that the genetic bias is intense enough to cause a substantial division of labor even in the most free and most egalitarian of future societies. [. . .] Even with identical education and equal access to all professions, men are likely to continue to play a disproportionate role in political life, business, and science" (Edward O. Wilson, quoted by Gould 1977, 259). For this third form of racism of destiny—completely detached from the racism of origins, since at this point all forms of race have disappeared from the horizon—one could perhaps choose the definition of *categorical racism.* A racism without races that highlights how Darwinian science irresistibly tends to transform itself into a bizarre metaphysics of the differential rates of biological potentiality, whose claims are to say the least questionable.

c. *The War of Races and the War of the Human Race.* To conclude, let us focus on a further implication of the racism of destiny, which pushes our metaphysical racism to its extreme consequences. Here, it is no longer a question of the past, and not even of the present: we are dealing with the future, ours, of *human beings to come.* Who are "we," who live in the world of autonomous life? We are the "living." This entails that wherever we turn our gaze, toward Darwin or Schelling, we are the *warriors of life.* In a well-known passage of the *Origin,* Darwin reminds us that he is using the expression "struggle for existence" in a "large and metaphorical sense," meaning that he is talking about a generalized struggle, both inter- and intra-specific, among living beings. This struggle, he explains, takes place in order to produce the greatest number of offspring (Darwin 2009b, 50), that is to say, in order to invade that which would later be defined as a "vital space." Such is the purposiveness without a purpose of the struggle for life that is fought by each single form-of-life. One could say something similar about Darwinian *war.* It also should be understood in a large and metaphorical sense. War is engaged in by individuals of different species and by individuals of the same species, which however belong to different varieties. In this last case one can speak of a war between races. War exists, at least in the case of natural selection, only where some win and others lose the battle for life. In this sense all those that win do not fight each other, in this strange world, since they don't triumph over each other. On the contrary, those that win belong to the same race, the one that demonstrates its superiority. Thus, in the world of autonomous life we have a state of permanent war, because it is Life itself that is warrior-like, and we have a permanent racism, because it is Life itself that is racist. Racism and war are the fixed qualities of this dramatic scenario; they are the two sides of the same coin, the two sides of the same metaphysical biology, with almost infinite variants. But what does this mean? It means that, if we remove the biological clothing from this argumentative framework, leaving only its metaphysical skeleton standing, at bottom each war of the races is always the same war, the only really permanent war, which nourishes the only equally permanent racism, beneath its many biological masks. Beneath it all, ultimately, we find the war of the human Race, which might also be called the war of human beings to come—*our war.* A metaphysical, total, eschatological war. A metaphysical war, because finally it is relieved of the burden of proposing a positive biological definition of the superior race: the human Race is a negative concept, an empty universal. An eschatological war, because it is projected toward the infinite conquest, never achieved, of the human as such: the human Race is the purposiveness without a purpose of a future that is regularly postponed. A total war, because it is engaged against all specifications of the human: the human Race, which coincides with human Life, must not bear the traces of any belonging. The political remainder of this war is our *freedom to live,* which unleashes a continual, paroxysmal search for our vital measures, to be now thought of as human measures, in that exhausting race for "perfect health" that some have defined as the "new utopia," the "Great Health" (Sfez 1995; Forti and Guaraldo 2006). And it is on this basis that the latest declension of the racism of destiny immediately takes hold: the racism of autonomous life, the racism of the freedom to live, the racism of the men to come. Who are we, the men of the "West"? We are not the men of the past; we do not recognize one another for what we have been. We are the men of the future who recognize one another for what they shall be. We are the men who will be human some day. We are the men of destiny, adults positioned to guard the human Race so that no one may any longer fill the empty universal, the purposiveness without a purpose, the lack of specific belonging of human Life; and for this reason, "we," *abstract race,* are in a state of permanent war against all the "others," against all *concrete races* that do not adhere to our religion. This is an invasive war, like all Darwinian wars, in the sense that it systematically tends to occupy the entire "vital space"; and it is a war for progeny, like all Darwinian wars, in the sense that it tends to systematically preserve us as the men to come. But it is a war only so long as some win and some lose, and for this reason it remains a war devoted to extermination and supremacy, like the preceding wars of races. The only difference is that, in this compass we defined as "metaphysical biology," the needle no longer points toward the second term but toward the first. In brief, ours is no longer a biological racism but a purely metaphysical one. Ours is an abstract race that opposes itself vigorously to all concrete races, to all human groups built around some preexisting qualification (mythical, religious, historical, or some other type) of its members. The human Race does not tolerate, ultimately, any form of prospectivism, nor any form of relativism, polytheism, pluralism, as far as the interpretation of the human is concerned. The human Race accepts, it does, the conflict of interpretations, but it also contemplates that only one of them will prevail over all the others. Once one has understood this, one can perhaps better understand the convergences and divergences between the old social Darwinism and a much more radical political Darwinism, in which the tomorrow "we" embody already shines.

d. *Social Darwinism and Political Darwinism.* Who are "we"? This question, that is the political question par excellence, has no meaning today. The only question that does have some meaning is the same one, but formulated in the future tense, no longer in the present. Who will "we" be? In this question mark, distorting the very geometry of our political space, the three axioms of autonomy glitter: the empty universal, the purposiveness without a purpose, and the lack of specific determination of the human Race. But these axioms can be treated differently depending on whether one gives a full or an empty definition of the human. One can take them for what they basically are, namely negative concepts that divide humans from humanity with the blade of life and will, thus abandoning humans to an impossible search, the search for their own humanity, which dissolves in the amorphous intensity, in the diaphanous selfhood of a peremptory *will-to-live*; or else, one can look at them the other way round, one can try to turn them into positive concepts that still stutter some vague and risky formula on the humanity of humans, thus attempting to outline some bio-anthropological theses that might eventually make sense of our history—a history otherwise doomed to run after an empty tomorrow that is the purposiveness without a purpose of natural History, the purposiveness without a purpose of the human Race. The former is a literal interpretation, the latter a metaphorical one of the identity between human form and vital force. The metaphorical interpretation is the one given by social Darwinism or, to use a definition more current today, by sociobiology. From this point of view, it is a matter of describing humankind—in all its species-specific aspects, which include the cultural, political, social, and psychic aspects of the human condition—in biological terms, or with the abstract laws of life, without totally erasing the very idea of human nature. Humans then become a biological species, and human beings are subdivided into superior and inferior races; or the differential threshold between strong and weak begins to pass elsewhere, between man and woman for instance. In any case, somewhere a threshold must exist, even in the most recent and polished versions of social Darwinism (Wilson 1978; Lopreato 1984), since there must exist somewhere a *differential* criterion to outline the biological appearance of the superior Race that better expresses our nature. Hence a regular pairing of reductionism and racism in a biologistic vein. But what about the literal interpretation of the three axioms of autonomy? This interpretation is, in reality, the keystone for all others, also for those that attempt to attenuate its disruptive abstraction, trying to fill the gap between human beings and human nature with the anthropological biologemes of sociobiology. This literal, no longer metaphorical, interpretation of the identity of human and life as such no longer attributes a recognizable appearance to the human "we"; it limits itself to recognizing a—pure and abstract—metaphysical appearance for it, that of the human Race, the Race of destiny. From this point of view, human beings are completely abandoned to their tomorrow, overwhelmed and lifted by a force, Life, that forbids its humanization *a priori* but commands it nonetheless. As a result, human beings are condemned to act out the impossible, i.e., becoming human. And they remain hanging on the projectuality of a History devoid of an ending. Can we speak of Darwinism in this regard? We can, and we can also take advantage from such definition, but only on condition of speaking of political Darwinism at this point, and no longer of social Darwinism. The advantage is that this definition casts some light on the close ties between our so-called ethnocentrism and our racism. Political Darwinism means that there is not one without the other, that our modern identity of "living beings" is only practicable in a war against others. Who are "we"? What is our ethnicity? It is no longer something cultural or historical, which belongs to our past, nor is it something that we can really describe in biological terms if, as we are now attempting, we distance ourselves from all variants of social Darwinism to change course towards political Darwinism. Today, we define ourselves in the future tense, no longer in the past. This is the time of "Western" man. So, what is our ethnicity? It is a metaphysical ethnicity; it is an invisible and untimely race; it is the Race of men to come, the Race of adults who are distinguished by an awareness they have acquired that their lineage does not come from the past, but from the future. This is the Race of "Western" destiny, which is the tomorrow of humans and life, the tomorrow, deprived of purpose, of their literal yet delusional identity, against whose backdrop an implacable metaphysical racism against all those who don't embrace our religion and insist on adoring their own idols, their own representations of the human, takes hold. *We are racists,* in our modern invisibility to ourselves, *because it is only in this manner that we manage to recognize ourselves among "us,"* employing an obscure, metaphysical criterion of distinction, that categorical differential between "us" and the "others" that is Life itself, captured in its propulsive abstraction, which dissolves all human form in the *purposiveness without a purpose* of human History. *We are racists,* and we don't even notice. Worse still, we are the last racists, *because we already embody the future, we embody human beings to come,* we embody the eschatological Race that by definition and on principle cannot be followed by any other. Our racism is for this reason no longer associated with a modest scientific reductionism but rather with an ambitious metaphysical eliminativism. All races that differ from ours become *trifling,* not worthy of notice, whatever we mean by different race. All human attributes, all finalities of human beings, all sense of belonging, can disappear in its insignificance, can be blown away by our battle-cry—*Salus vitae.* We inhabit these extremities, these blanks, which bind us together and make the difference between "us" and the "others," although we remain *non-apparent and completely imperceptible to ourselves.* We human beings who have managed, in one and the same stroke, with a powerful metaphysical gesture, to both invent and vanify our own humanity, we are the last. *Ultimi barbarorum.*

| | | |

A specter is haunting the world. One could give it several names, and certainly it would be advisable to go over its multiple facets. But why do I insist on calling it Darwinism? Because of two interrelated motives: first because the circle of autonomy closes into a metaphysics of *life itself,* which is the invisible cipher of our world; second, because Darwin more than anyone else traced the north and the south, the east and the west of this metaphysical world, which is, precisely, the world in which we *live.* It is therefore little use to make surreptitious distinctions between Darwin and social Darwinism, arguing that the same ideas, like the war between races or the extermination of the weakest, could have been elaborated without even mentioning Darwin (La Vergata 2005), since in actual fact it was in Darwin's metaphysical language that these ideas were formulated, and, if we look more deeply, beneath the veil of social Darwinism we can perceive the specter of a political Darwinism, which runs around the world wearing ever-changing masks (today's "political realism" is one of these), and it is always ready to provide sustenance to the worst nightmares of social Darwinism, with new colors and perhaps new words, as is the case with contemporary sociobiology. Darwinism—seen as the metonymical figure of something larger, which includes and simultaneously exceeds the edges of the Darwinian speculative framework—is the hole, the funnel into which the metaphysics of autonomy after Kant has precipitated, only to come out of it welded to a series of assertions that we still consider to be true, not so much on the basis of clear scientific evidence, which on the contrary tend to challenge all current understandings of biological phenomena (Continenza and Gagliasso, 1996) as by virtue of a metaphysical hallucination whose reasons need to be carefully analyzed. *Noli tangere Darwin.* It is difficult not to notice that this taboo exists in the "Western" world. We live beneath the shade of a kind of Darwinian dogma on which it would be worthwhile to reflect. Is there anyone, for instance, who dares think that the hypothesis of natural selection will sooner or later be abandoned in favor of a further hypothesis? Is there anyone who dares do this without immediately feeling obliged to attribute the paternity of a possible "reinvention" of biology to Darwin himself, looking for it in the *Notebooks* or elsewhere (Eldredge 2005), almost as if Darwinism were a knowledge apparatus similar to Marxism or Freudism, in which Foucault taught us to weigh the function performed by the founding father's words (Foucault 1998)? And yet an impudent and shameless innovation is what science boasts of: why should biology be immune? Why shouldn't it cast off the hypothesis of natural selection one day? Why do we give Darwin such enormous credit, metaphysical more than scientific, equal only to that which in the past we reserved for the names of Aristotle, Thomas Aquinas, Descartes (and a few others)? In the final analysis, the answer lies in the theorem that modernity is hinged on, the *theorem of the liar.* But to articulate this theorem, we first need to recapitulate the reasons Darwinism is not only science but also metaphysics. Basically, the reasons are three, and they read as follows:

a. *The Origin of Species* is a book to which we must attribute the status and dignity of a classic of metaphysics, since it is a book that presented a theory that at the time was devoid of plausible scientific foundations and that, for most people, will become credible only with the so-called new synthesis in the 1930s and 1940s.

b. *Darwinian sociology* is a complex of metaphysical theories whose incandescent nucleus can be found in Darwin's own texts. The very idea that lies behind the disciplinary trespassing from biology into sociology is metaphysical and not scientific. This is the idea of biological reductionism according to which all specific human traits (psychic, social, cultural) ought to be described and explained in biological terms. Thanks to this metaphysical move, sociologists, for over a century now, satisfy their "desire to be *scientists*" (Rose, Lewontin and Kamin 1983, 238).
c. *Darwinian biology* is a set of scientific theories, gathered however under a lowest common denominator, beneath a sort of metaphysical indicator (Foucault would have called it an "epistemological indicator") that is the abstract, yet dense and geometrically constructed, notion of Life as such. In this cryptometaphysical notion a new ontology is hidden, which is an ontology of the Self and no longer an ontology of the One. The Self is the secret support of natural selection (as clearly attested by the theory of the *selfish gene*).

The thesis I am setting out here is that Darwinism is a metaphysics in all three senses, *since, further upstream, the metaphysical is the abstraction of autonomous life.* Seen in this light, the Darwinian theory of evolution, centered around the principle of natural selection, is not at all tautological, as is sometimes objected, since it is instead functional to a precise definition of Life as such—a definition that is also shared by other currents of modern thought, but which finds its most powerful articulation in contemporary biology. This articulation, tendentially invisible, becomes obliquely, more evident by inserting Darwinian theory into its ontological hinterland, which is the metaphysical hinterland of "today," of "the present," of modern times. We, the moderns, are obsessed by a problem. How can we stitch the tear between humans and humanity back together? How can we make human beings into *sincere and trustworthy* representatives of humanity? What are the conditions of possibility of this operation, which is the most difficult but also the most imperative, the most cogent, operation modernity imposes on us?

These are indeed the queries unleashed by the (Kantian) metaphysical theorem of "today," of "the present," of modern times: human beings infallibly lie to themselves, and yet humans cannot always lie to themselves, to their humanity, without their being eclipsed into an absolute lie that would annihilate their very existence. Hence the need for constructing the conditions of possibility of a lie that would not eclipse humans completely, but safeguard their sincerity, displacing it over time, in the course of History, shoving it into the past or the future, depending on which metaphysical option one adopts. Labor-force and force-of-life thus become the expression of the ineffable autonomy of modern humans, within a syntactic and semantic grid that goes to make up modernity's unconscious, with its two metaphysical reservoirs of sense, economics and biology, whose statements are valid not so much because of their truth but because of their strange appearance of truthfulness. If human beings can be truthful "today," they cannot in any case be so by overcoming their intrinsic opacity; they can only be so without being so, insofar as they are mesmerized by a metaphysical hallucination, the hallucination of History, which traces the conditions of possibility of their lying and not lying, simultaneously, ubiquitously, to themselves. Trapped in this paradox, modern humans, who once worked and now prefer to live, end up being forced into acting out an impossibility, that of humanizing themselves, while still finding a source of *sincere self-deceit* in this acting out. That, in a few words, is the metaphysical hallucination of "today."

Is it possible to escape it? Is it possible to exit modernity? Is it possible to move beyond "the present," which, in the name of an irreducible opacity of human beings, tends to expand its disruptive void and exasperate its unstoppable *power*? If it were possible, this would amount to sedating modernity. Instead of eliminating its constitutive opacity, we should indeed try to symbolize it, to write it, to make it into an insurmountable anchor point that one can no longer dream of overcoming precisely because one has finally written the cipher of an impossibility into this overcoming. But what does this all mean? First of all it means that the exit from modernity cannot be a historical exit; it can only be a metaphysical exit, because the times in which we live are eminently metaphysical, although invisible to themselves. Secondly it means that the exit from modernity seems to announce itself in the form of a *sedated metaphysics* that, instead of running after the mirage of actualizing modernity's power, does no more than extinguish hallucination.

| | | |

Finding the right door to exit from modernity is an exigency that has been felt for a long time, from the time modernity irrupted among "us." People have tried to open many doors, without this yet having allowed us to reach the result we yearned for. The closest doors, at the moment, seem to be three.

a. The first door is that which takes us back to old religions. In spite of appearances, this door does not open onto Paradise. To which ancient God can "Western" human beings return? Should we listen to Schelling, to a two-faced God, a God with a light face, the ancient face of transcendence, and a dark face, that of God's foundation and reason (*Grund*) in a world that has since changed—the face of the new and ferocious divinity of Life? This is what Schelling showed us in his visionary rigor, namely that the God of ancient religions can be born again, in our world, but that God, however, tends to do so obeying the imperatives of a new divinity, which upsets the old order and gives birth to diabolical forms of religiosity, cross-eyed forms of devotion to a God just as cross-eyed and schizoid, who has forgotten the very meaning of the word God. In this sense the current problem of religion is not that of coming to terms with modernity, it is that of not *excessively* coming to terms with it. Only on this condition could religion again take on the role of metaphysical vanguard. But this remains a remote possibility after all. Let us burn our fingers with a sensitive example: the sacralization of the fetus from conception, of a fetus in which we all perceive the peremptory expression of a human Life, was not always part of the Church's teachings; it is already a concession made by the old religion to the new idol of modernity. And this is not only because the fetus was not something visible until a few decades ago, but above all because the "substantive life," the Life hypostasized in the image of the fetus, is a metaphysical hallucination of our times, which has very little in common with the revealed Life of the Christian God (John 11:25). In the name of this metaphysical hallucination, nevertheless, the imperative value of a faceless and abstract Life is given priority over a human face, the real face of a woman, on the basis of categorical certainties that are taken for granted and presented as scientific (Duden 1993). Although it might seem strange at first, it is also from an example like this that one can infer what kind of pressure modern metaphysics exerts on our auscultation of the past, deforming (if not actually uprooting) our cultural identity.

b. The second door is that of a further advancement along the path of autonomous life that would proceed so far as to reconfigure, almost overturn, the semantics of "bare life," *bloßes Leben,* detaching it from a metaphysics of power. It is a door that has been jimmied repeatedly, but it is a door that no one so far has managed to fully open. We already discussed this option when talking about Canguilhem, and numerous other examples could be added—for instance the whole current of Bergsonism or, more generally, all philosophies that drew inspiration from "the fact," as abstract as it was oft repeated, "of life's willing itself" (Jonas 2001, 61). In lieu of a comment, it would suffice to repeat, each time, that it is difficult to think of exiting modernity, or of halting its power, by reckoning one could traverse its hallucinations unscathed. Having said this, it is worth noting that some contemporary philosophy, by recovering the category of "bare life" from Walter Benjamin's texts (who used it in a *critical, polemical* fashion, and in clear opposition to his times), today attempts to throw a wrench into the machine of modernity by using the idea of another "life" as lever: no longer Life but *a* life; no longer the empty universal but the "whatever-singularity"; no longer the abstract but the concrete, which is almost ineffable in its own concreteness. This attempt, with its neoexistentialist accents, is interesting because of two lessons that can be drawn from it, one negative the other positive. The first lesson is that, following this path while being aware of the traps of modern "life," inhabited by deadly and impending drives, one symptomatically ends up being dumbstruck. We detach ourselves from ourselves, from the words we use and breathe since birth, without reaching a different land. Poetic allusion can, then, take the place of critical argument. The second lesson is that, once words have vanished, one ends up feeling that *exiting modernity* is an illusory task. It is precisely an imperative of this kind, *exiting modernity,* that makes us modern and makes us, ultimately, so *powerful.*

c. The third door to exit modernity does not therefore take us out of modernity, it leads us inside: it is a door that closes, instead of opening, sealing the confines of our identity. In this way we can detach the "today" from tomorrow; we can decide and opt for our finitude, by tracing a threshold that restores contingency to historical time. This is what is usually described as the critical function of philosophy, a function that is also exquisitely modern. But modernity is in continuous, blatant contradiction with itself and allows us to take sides. On the one hand, we can *plan* to heal the contradiction that we are, thus hallucinating a History that we are expected to achieve and to realize. On the other, we can *criticize* ourselves; we can inhabit the contradiction that we are, thus easing the burden of History on our historical condition. We could call them the hallucinatory and the symptomatic paths of modernity. The difference between them is not that one is true and the other is false, because truth cannot be reached anywhere in this case. It is rather a matter of awareness. We are an enigma for ourselves, an enigma that can be treated and sedated only by means of a contingent writing, which is not devoid of political valence. Allowing the enigma that we are to be written—*we, the moderns, live*—does in fact offer us the opportunity to listen to ourselves, in a sort of criticism squared, or in a sort of metacriticism of modernity's own criticism. And such a listening to ourselves is the first step, a step backward, in order to listen to others, who are staring at us, in their indestructible fragility.

TRANSLATOR'S ACKNOWLEDGMENTS

I WOULD NOT HAVE BEEN INVOLVED in this translation project without the help of and contacts established by Andrea Righi, colleague and friend. The project combined two of my major areas of interest, philosophy and the natural sciences (and the history of science). As with a prior project whose author, Alessandro Minelli, has become a dear friend, the intense exchanges and in-depth help and assistance provided by this author, Davide Tarizzo, especially in the area of preferred English translations, led first to a deepening dialogue and now to a friendship based on shared curiosities and interests in a number of fields. I would especially like to thank Erin Warholm-Wohlenhaus of the University of Minnesota Press for her encouragement, constant attention to the work-in-progress, and assistance with many editorial and practical matters, several of which she had no direct responsibility for: a care and attention well above and beyond the "call of duty."

BIBLIOGRAPHY

Agamben, Giorgio. 1998. *Homo Sacer: Sovereign Power and Bare Life*. Stanford, Calif.: Stanford University Press.

Arendt, Hannah. 1958. *The Human Condition*. Chicago: University of Chicago Press.

———. 1961. *Between Past and Future. Six Exercises in Political Thought*. New York: Viking.

———. 1963a. *On Revolution*. New York: Viking.

———. 1963b. *Eichmann in Jerusalem. A Report on the Banality of Evil*. New York: Viking.

———. 1968. "Truth and Politics," in *Between Past and Future: Eight Exercises in Political Thought*. New York: Viking.

———. 1972. "Lying in Politics. Reflections on the Pentagon Papers," in *Crises of the Republic*. New York: Harcourt, Brace & Jovanovich.

———. 1978. *The Life of the Mind*. San Diego: Harcourt, Brace & Jovanovich.

———. 1982. *Lectures on Kant's Political Philosophy*. Chicago: University of Chicago Press.

———. 1993. *Was ist Politik?* Munich: Piper.

Badiou, Alain. 2007. *The Century*. Cambridge: Polity Press.

Barrow, John D., and Frank J. Tripler. 1986. *The Anthropic Cosmological Principle*. Oxford: Oxford University Press.

Barrow, John D., Simon Conway Morris, Stephen Freeland, and Charles L. Harper, eds. 2007. *Fitness of the Cosmos for Life*. Cambridge: Cambridge University Press.

Beatty, John. 1992. "Fitness: Theoretical Contexts," in *Keywords in Evolutionary Biology*, ed. Evelyn Fox Keller and Elisabeth A. Lloyd. Cambridge, Mass.: Harvard University Press.

Behe, Michael J. 2006. *Darwin's Black Box: The Biochemical Challenge to Evolution,* 2nd ed. New York: Free Press.

Benjamin, Walter. 1996. "Critique of Violence," in *Selected Writings,* ed. Marcus Bullock and Michael W. Jennings. Cambridge, Mass.: Harvard University Press. [Also in *Reflections,* Schocken, 1986]

Berlin, Isaiah. 1969. *Four Essays on Liberty*. Oxford: Oxford University Press.

———. 1978. *Concepts and Categories*. London: Hogarth Press.

Bichat, Xavier. 1978. *Physiological Researches on Life and Death*. Washington, D.C.: University Publications of America.

Blumenbach, Johann. 1971. *Über den Bildungstrieb und das Zeugungsgeschäfte*. Stuttgart: Gustav Fischer. [English translation, online source: http://galenet.galegroup.com/servlet/ECCECCO?c=1&stp=Author&ste=11&af=BN&ae=T061381&tiPG=1&dd=0&dc=flc&docNum=CW109805269&vrsn=1.0&srchtp=a&d4=0.33&n=10&SU=0LRM&locID=prin77918. London: printed for T. Cadell, 1792.]

Bobbio, Norberto. 1996. *The Age of Rights*. Cambridge, Mass.: Polity Press.

Bodei, Remo. 1991. *Geometria delle passioni. Paura, speranza, felicità: filosofia e uso politico*. Milan: Feltrinelli.

Bori, Pier Cesare. 2000. *Pluralità delle vie. Alle origini del discorso di Pico della Mirandola*. Milan: Feltrinelli.

Bowler, Peter J. 1984. *Evolution: The History of an Idea*. Berkeley: University of California Press.

Brandon, Robert N. 1978. "Adaptation and Evolutionary Theory," *Studies in the History and Philosophy of Science* 9, no. 3:181–206.

———. 1992. "Environment," in *Keywords in Evolutionary Biology,* ed. Evelyn Fox Keller and Elisabeth A. Lloyd. Cambridge, Mass.: Harvard University Press, 1992.

Buffon, Georges-Louis Leclerc de. 1765. *Buffon's Natural History*. London: Sold by H.D. Symonds.

Burian, Richard M. 1992. "Adaptation: Historical Perspectives," in *Keywords in Evolutionary Biology,* ed. Evelyn Fox Keller and Elisabeth A. Lloyd. Cambridge, Mass.: Harvard University Press.

Camardi, Giovanni. 2001. "Richard Owen: Morphology and Evolution," *Journal of the History of Biology* 34, no. 3:481–515.

Canguilhem, Georges. 1955. *La formation du concept de réflexe aux XVII et XVIII siècles*. Paris: Presses Universitaires de France.
———. 1965. *La connaissance de la vie*. Paris: Vrin.
———. 1968. *Etudes d'histoire et de philosophie des sciences concernant le vivant*. Paris: Vrin.
———. 1990. "Le cerveau et la pensée," in *Georges Canguilhem: Philosophe, historien des sciences*. Paris: Albin Michel.
———. 1991. *On the Normal and the Pathological*. New York: Zone Books.
———. 2000. "Sur l'histoire des sciences de la vie depuis Charles Darwin," in *Idéologie et rationalité dans l'histoire des sciences de la vie*. Paris: Vrin.
———. 2012. *Writings on Medicine*. New York: Fordham University Press.
Cavarero, Adriana. 2009. *Horrorism: Naming Contemporary Violence*. New York: Columbia University Press.
Cesa, Claudio. 1994. *Introduzione a Fichte*. Rome-Bari: Laterza.
Chamberlain, Houston Stewart. 1912. *Wehr und Gegenwehr*. Munich: Bruckmann.
Charlton, William. 1988. *Weakness of Will: A Philosophical Introduction*. Oxford: Blackwell.
Coleridge, Samuel Taylor. 1848. *Hints Towards the Formation of a More Comprehensive Theory of Life*. London: Churchill.
Colwell, Robert K. 1992. "Niche: A Bifurcation in the Conceptual Lineage of a Term," in *Keywords in Evolutionary Biology*, ed. Evelyn Fox Keller and Elisabeth A. Lloyd. Cambridge, Mass.: Harvard University Press.
Continenza, Barbara, and Elena Gagliasso. 1996. *Giochi aperti in biologia: Una riflessione critica su adattamento, struttura, specie*. Milan: Franco Angeli.
Cuvier, Georges-Louis. 1817. *Le règne animal*. Paris: Belin.
Darwin, Charles. 1896. *The Variation of Animals and Plants under Domestication*. New York: Appleton.
———. 1909. *Sketch*. Cambridge: Cambridge University Press.
———. 1987. *Notebooks 1836–1844*. Cambridge: Cambridge University Press.
———. 1996. *Letters: A Selection, 1825–1859*. Cambridge: Cambridge University Press.

———. 2005. *The Autobiography of Charles Darwin*. New York: Barnes & Noble.

———. 2009a. *The Descent of Man*. New York: Cambridge University Press.

———. 2009b. *On the Origin of Species by Means of Natural Selection, or the Preservation of Favoured Races in the Struggle for Life*. New York: Cambridge University Press.

Darwin, Francis. 1959. *The Life and Letters of Charles Darwin. Volume Two*. New York: Basic Books.

Dawkins, Richard. 1982. *The Extended Phenotype: The Long Reach of the Gene*. Oxford: Oxford University Press.

———. 2006. *The Selfish Gene,* Thirtieth Anniversary edition. Oxford: Oxford University Press.

Deleuze, Gilles, and Félix Guattari. 1983. *Anti-Oedipus: Capitalism and Schizophrenia*. Minneapolis: University of Minnesota Press.

Dennett, Daniel C. 1978. *Brainstorms: Philosophical Essays on Mind and Psychology*. Cambridge, Mass.: MIT Press.

———. 1987. *The Intentional Stance*. Cambridge, Mass.: MIT Press.

———. 1991. *Consciousness Explained*. Boston: Little, Brown & Co.

———. 1996. *Darwin's Dangerous Idea: Evolution and the Meanings of Life*. New York: Simon & Schuster.

———. 2003. *Freedom Evolves*. New York: Viking.

Descartes, Rene. 2010–2015. *The Passions of the Soul,* ed. Jonathan Bennett, http://www.earlymoderntexts.com/assets/pdfs/descartes1649part2.pdf.

Dickinson, Robert E. 1969. *The Makers of Modern Geography*. London: Routledge.

Diderot, Denis. 1999. *Thoughts on the Interpretation of Nature, and Other Philosophical Works*. Manchester, UK: Clinamen Press.

Dihle, Albrecht. 1982. *The Theory of Will in Classical Antiquity*. Berkeley: University of California Press.

Dobzhansky, Theodosius. 1937. *Genetics and the Origin of Species*. New York: Columbia University Press.

Dodds, Eric D. 1951. *The Greeks and the Irrational*. Berkeley and Los Angeles: University of California Press.

Duden, Barbara. 1993. *Disembodying Women: Perspectives on Pregnancy and the Unborn*. Cambridge, Mass.: Harvard University Press.

———. 2002. *Die Gene im Kopf—der Fötus im Bauch: Historisches zum Frauenkörper*. Hannover: Offizin Verlag.

Dumont, Louis. 1977. *Homo aequalis I: Genèse et épanouissement de l'idéologie économique*. Paris: Gallimard.

———. 1986. *Essays on Individualism: Modern Ideology in Anthropological Perspective*. Chicago: University of Chicago Press.

———. 1991. *Homo aequalis II. Idéologies nationales comparées: l'idéologie allemande*. Paris: Gallimard.

Duris, Pascal, and Gabriel Gohau. 1997. *Histoire des sciences de la vie*. Paris: Nathan.

Dworkin, Ronald M. 1993. *Life's Dominion: An Argument about Abortion and Euthanasia*. New York: Harper Collins.

Eldredge, Niles. 1995. *Reinventing Darwin: The Great Debate at the High Table of Evolutionary Theory*. New York: John Wiley & Sons.

———. 2005. *Darwin: Discovering the Tree of Life*. New York: Norton.

Esposito, Roberto. 2008. *Bíos: Biopolitics and Philosophy*. Minneapolis: University of Minnesota Press.

———. 2010. *Communitas: The Origin and Destiny of Community*. Stanford, Calif.: Stanford University Press.

———. 2011. *Immunitas: The Protection and Negations of Life*. Cambridge: Polity Press.

Fichte, Johann Gottlieb. 1959. *Versuch einer neuen Darstellung der Wissenschaftslehre*, trans. into Italian as *Teoria della scienza 1798 "Nova Methodo,"* Milan-Varese: Istituto Editoriale Cisalpino.

———. 1982. *Science of Knowledge*. Cambridge: Cambridge University Press. [online: https://voices.uchicago.edu/germanphilosophy/files/2012/05/Fichte-The-Science-of-Knowledge-sec-1–3.pdf.]

———. 2000. *Foundations of Natural Right*. Cambridge: Cambridge University Press.

———. 2005. *The System of Ethics*. Cambridge: Cambridge University Press.

Fimiani, Mariapaola. 1997. *Foucault e Kant. Critica clinica etica*. Naples: Città del Sole.

Fisher, Ronald A. 1930. *The Genetical Theory of Natural Selection*. Oxford: Clarendon Press.

———. 1934. "Indeterminism and Natural Selection," *Philosophy of Science* 1:99–117.

Flanagan, Owen. 2003. "Ethical Expressions: Why Moralists Scowl, Frown and Smile," in *The Cambridge Companion to Darwin,* ed. Jonathan Hodge and Gregory Radick. Cambridge: Cambridge University Press.

Forti, Simona, and Olivia Guaraldo. 2006. "Rinforzare la specie: Il corpo femminile tra biopolitica e religione materna," *Filosofia politica*. 1:57–78.

Foucault, Michel. 1978. *The Will to Knowledge: History of Sexuality, Volume I*. New York: Pantheon.

———. 1979. "Cuvier's Position in the History of Biology," translated by Felicity Edholm. *Critique of Anthropology* 4:13–14 (Summer): 125–30.

———. 1984. "What Is Enlightenment?" trans. Catherine Porter, in *The Foucault Reader,* ed. P. Rabinow, 32–50. New York: Pantheon.

———. 1985. *Discourse and Truth: The Problematization of Parrhesia*. Evanston, Ill.: Northwestern University Press. [also at: http://foucault.info/documents/parrhesia/.]

———. 1998. "What Is an Author?" in *Aesthetics, Method, and Epistemology,* vol. 2, ed. James D. Faubion. New York: New Press.

———. 2003a. "The Birth of Biopolitics," in *The Essential Foucault,* ed. P. Rabinow and N. Rose, 202–7. New York: New Press.

———. 2003b. *"Society Must Be Defended": Lectures at the Collège de France, 1975–1976*. New York: Picador.

———. 2005. *The Order of Things*. London: Routledge; e-book Taylor & Francis E-Library.

Foucault, Michel, and Noam Chomsky. 2006. *The Chomsky–Foucault Debate: On Human Nature*. New York: New Press.

Fox Keller, Evelyn. 1992. "Fitness: Reproductive Ambiguities," in *Keywords in Evolutionary Biology* ed. Evelyn Fox Keller and Elisabeth A. Lloyd. Cambridge, Mass.: Harvard University Press.

Fredrickson, George M. 2002. *Racism: A Short History*. Princeton, N.J.: Princeton University Press.

Freud, Sigmund. 1961. *Beyond the Pleasure Principle*. New York: Norton.

Galli, Carlo. 2010. *Political Spaces and Global War*. Minneapolis: University of Minnesota Press.

Gasman, Daniel. 1971. *The Scientific Origins of National Socialism: Social*

Darwinism in Ernst Haeckel and the German Monist League. London: MacDonald.

———. 1998. *Haeckel's Monism and the Birth of Fascist Ideology*. New York: Peter Lang.

Gauchet, Marcel. 1997. *The Disenchantment of the World: A Political History of Religion*. Princeton, N.J.: Princeton University Press.

Gayon, Jean. 1998. *Darwinism's Struggle for Survival: Heredity and the Hypothesis of Natural Selection*. Cambridge: Cambridge University Press.

———. 2003. "From Darwin to Today in Evolutionary Biology," in *The Cambridge Companion to Darwin*, ed. Jonathan Hodge and Gregory Radick. Cambridge: Cambridge University Press.

Gehlen, Arnold. 1983. *Philosophische Anthropologie und Handlungslehre*. Frankfurt a.M.: Klostermann.

———. 1988. *Man, His Nature, and Place in the World*. New York: Columbia University Press.

Gilson, Étienne. 1984. *From Aristotle to Darwin and Back Again: A Journey in Final Causality, Species, and Evolution*. Notre Dame: Indiana University Press.

Gliboff, Sander. 1998. "Evolution, Revolution, and Reform in Vienna: Franz Unger's Ideas on Descent and Their Post-1848 Reception," *Journal of the History of Biology* 31.

Goethe, Johann Wolfgang. 1988. *Scientific Studies*. New York: Suhrkamp.

Gould, Stephen Jay. 1977. *Ever Since Darwin: Reflections in Natural History*. New York: Norton & Co.

———. 1996. *Dinosaur in a Haystack: Reflections in Natural History*. New York: Three Rivers Press.

———. 2002. *The Structure of Evolutionary Theory*. Cambridge, Mass.: Harvard University Press.

Gould, Stephen Jay, and Richard Lewontin. 1979. "The Spandrels of San Marco and the Panglossian Paradigm," in *Proceedings of the Royal Society of London*, B 205.

Griesemer, James R. 1992. "Niche: Historical Perspectives," in *Keywords in Evolutionary Biology*, ed. Evelyn Fox Keller and Elisabeth A. Lloyd. Cambridge, Mass.: Harvard University Press.

Habermas, Jürgen. 1994. "Struggles for Recognition in the Democratic Constitutional State," in *Multiculturalism: Examining the Politics of Recognition,* ed. Charles Taylor et al. Princeton, N.J.: Princeton University Press.

———. 1998. *The Inclusion of the Other: Studies in Political Theory.* Cambridge, Mass.: MIT Press.

Hadot, Pierre. 2004. *La voile d'Isis: Essai sur l'histoire de l'idée de nature.* Paris: Gallimard. [*The Veil of Isis: an Essay on the History of the Idea of Nature.* Cambridge, Mass.: Harvard University Press, 2006.]

Haeckel, Ernst. 1934. *Riddle of the Universe.* London: Watts & Co.

Hare, Richard M. 1965. *Freedom and Reason.* Oxford: Oxford University Press.

Hawkins, Mike. 1997. *Social Darwinism in European and American Thought, 1860–1945.* Cambridge: Cambridge University Press.

Hegel, Georg Wilhelm Friedrich. 1977. *Phenomenonology of Spirit.* Oxford: Oxford University Press.

———. 1991. *Elements of the Philosophy of Right.* Cambridge: Cambridge University Press.

———. 2001. *The Phenomenology of Mind.* Blackmask Online. http://www.naturalthinker.net/trl/texts/Hegel,G.W.F/Hegel,_G.W.F._-_The_Phenomenology_Of_Mind.pdf.

———. 2010. *The Science of Logic.* Cambridge: Cambridge University Press.

Heidegger, Martin. 1983. *Die Selbstbeahauptung der deutschen Universität.* Frankfurt a. M.: Klostermann. [English translation available at: http://la.utexas.edu/users/hcleaver/330T/350kPEEHeidegger Self-Assertion.pdf .]

———. 1991. *Nietzsche,* vols. 3–4. New York: HarperCollins.

———. 1996. *Being and Time.* Albany: State University of New York Press.

———. 1997. *Kant and the Problem of Metaphysics.* Bloomington: Indiana University Press.

———. 2003. *The End of Philosophy.* Chicago: University of Chicago Press.

Henderson, Lawrence J. 1913. *The Fitness of Environment: An Inquiry into the Biological Significance of the Properties of Matter.* New York: Macmillan.

Herder, Johann Gottfried. 1966. *Outlines of a Philosophy of the History of Man*. New York: Bergman.

Herrnstein, Richard J., and Charles Murray. 1994. *The Bell Curve: Intelligence and Class Structure in American Life*. New York: Free Press.

Himmelfarb, Gertrude. 1959. *Darwin and the Darwinian Revolution*. London: Chatto & Windus.

Hitler, Adolf. 1939. *Mein Kampf*. Boston: Houghton Mifflin Company.

Hofstadter, Richard. 1945. *Social Darwinism in American Thought, 1860–1915*. Philadelphia: University of Pennsylvania Press.

Hume, David. 1996. *Enquiries Concerning Human Understanding and Concerning the Principles of Morals*. Oxford: Clarendon Press.

———. 2010. *An Enquiry Concerning the Principles of Morals*. Project Gutenberg ebook, #4320, released January 12.

Huntington, Samuel P. 1996. *The Clash of Civilizations and the Remaking of World Order*. New York: Simon & Schuster.

Hutchinson, George Evelyn. 1957. "Concluding Remarks," *Cold Spring Harbor Symposium on Quantitative Biology* 22.

Jacob, François. 1993. *The Logic of Life: A History of Heredity*. Princeton, N.J.: Princeton University Press.

Jankélévitch, Vladimir. 1957. *Le Je-ne-sais-quoi et le Presque-rien*. Paris: Seuil.

Jonas, Hans. 1979 *Das Prinzip Verantwortung: Versuch einer Ethik für die technologische Zivilisation*. Frankfurt a. M.: Insel.

———. 2001. *The Phenomenon of Life: Toward a Philosophical Biology*. Evanston, Ill.: Northwestern University Press.

———. 2003. "Essere e dover essere," *Micromega* 5.

Kant, Immanuel. 1977. *Über ein vermeintes Recht aus Menschenliebe zu lügen, in Werke in zwölf Bänden*, vol. 8. Frankfurt a. M.: Suhrkamp.

———. 1979. *The Conflict of the Faculties*. New York: Abaris.

———. 1987. *Critique of Judgment*. Indianapolis: Hackett.

———. 1991. *The Metaphysics of Morals*. Cambridge: Cambridge University Press.

———. 1996a. *Critique of Pure Reason*. Indianapolis: Hackett.

———. 1996b. *Anthropology from a Pragmatic Point of View*. Carbondale: Southern Illinois University Press.

———. 1997. *Lectures on Ethics*. New York: Cambridge University Press.

———. 2002a. *Groundwork for the Metaphysics of Morals*. New Haven, Conn.: Yale University Press.

———. 2002b. *Critique of Practical Reason*. Indianapolis: Hackett.

———. 2009. *Religion within the Bounds of Bare Reason*. Indianapolis: Hackett.

———. 2015. *On Perpetual Peace*. Peterborough, ON: Broadview Press.

Kellogg, Victor L. 1907. *Darwinism Today*. New York: Holt.

Kitcher, Philip. 2003. "Giving Darwin His Due," in *The Cambridge Companion to Darwin*, ed. Jonathan Hodge and Gregory Radick. Cambridge: Cambridge University Press.

Lacoue-Labarthe, Philippe, and Jean-Luc Nancy. 1991. *Le mythe nazi*. Paris: L'Aube.

Lakoff, George. 1988. "Cognitive Semantics," in *Meaning and Mental Representations*, ed. Umberto Eco, Marco Santambrogio, and Patrizia Violi, 119–54. Bloomington, Indiana University Press.

Lakoff, George, and Mark Johnson. 1999. *Philosophy in the Flesh: The Embodied Mind and Its Challenge to Western Thought*. New York: Basic Books.

Lamarck, Jean-Baptiste. 1984. *Zoological Philosophy: An Exposition with Regard to the Natural History of Animals*. Chicago: University of Chicago Press.

La Vergata, Antonello. 1990. *L'equilibrio e la guerra della natura: Dalla teologia naturale al darwinismo*. Naples: Morano.

———. 2005. *Guerra e darwinismo sociale*. Soveria Mannelli: Rubettino.

Lennox, James G. 1992. "Teleology," in *Keywords in Evolutionary Biology*, ed. Evelyn Fox Keller and Elisabeth A. Lloyd. Cambridge, Mass.: Harvard University Press.

Lenoir, Timothy. 1983. *Strategy of Life: Vital Materialism in Nineteenth-Century German Physiology*. Dordrecht: Reidel.

Lewontin, Richard. 1978. "Adaptation," *Scientific American*, 239.

Linné [Linnaeus], Carl von. 1767/1770. *Systema naturae, per regna tria naturae*. Vindobonae: Jo. Th. Trattern.

Lopreato, Joseph. 1984. *Human Nature and Bio-Cultural Evolution*. Boston: Allen & Unwin.

Lovejoy, Arthur O. 1936. *The Great Chain of Being: A Study of the History of an Idea*. Cambridge, Mass.: Harvard University Press.

Lukács, György. 1981. *The Destruction of Reason*. Atlantic Highlands, N.J.: Humanities Press.

Malthus, Thomas Robert. 1826. *An Essay on the Principle of Population*, 6th ed. London: Murray.

Marconi, Diego. 2001. *Filosofia e scienza cognitiva*. Rome-Bari: Laterza.

Maupertuis, Pierre-Louis Moreau de. 2005. *Essai sur la formation des corps organisés*. In Denis Diderot, *Pensées sur l'interpretation de la nature*. Paris: Flammarion.

Maynard Smith, John. 1986. *The Problems of Biology*. Oxford: Oxford University Press.

Mayr, Ernst. 1961. "Cause and Effect in Biology," *Science* 134.

———. 1982. *The Growth of Biological Thought, Diversity, Evolution, and Inheritance*. Cambridge, Mass.: Harvard University Press.

———. 1991. *One Long Argument: Charles Darwin and the Genesis of Modern Evolutionary Thought*. Cambridge, Mass.: Harvard University Press.

Moiso, Francesco. 1990. *Vita natura libertà: Schelling (1795–1809)*. Milan: Mursia.

Momigliano, Arnaldo. 1996. *Pace e libertà nel mondo antico: Lezioni a Cambridge gennaio-marzo 1940*. Florence: La Nuova Italia.

Monod, Jacques. 1971. *Chance and Necessity: An Essay on the Natural Philosophy of Modern Biology*. New York: Knopf.

Morange, Michel. 2008. *Life Explained*. New Haven, Conn.: Yale University Press.

Morselli, Enrico. 1904. *Introduzione e aggiunte*, in Ernst Haeckel, *I problemi dell'universo*, Turin: UTET.

Mosse, George L. 1978. *Toward the Final Solution: A History of European Racism*. New York: Howard Fertig.

Nietzsche, Friedrich. 2006. *Thus Spoke Zarathustra: A Book for All and None*. Cambridge: Cambridge University Press.

———. 2011. *Dawn: Thoughts on the Presumptions of Morality*. Stanford, Calif.: Stanford University Press.

Ospovat, Dov. 1979. "Darwin after Malthus," *Journal of the History of Biology* 12.

———. 1980. "God and Natural Selection: The Darwinian Idea of Design," *Journal of the History of Biology* 13.

Owen, Richard. 1992. *The Hunterian Lectures in Comparative Anatomy. May–June 1837*. London: Natural History Museum Publications.

Pearson, Karl. 1892. *Grammar of Science*. London: Scott.

Pichot, André. 1993. *Histoire de la notion de vie*. Paris: Gallimard.

———. 1999. *Histoire de la notion de gene*. Paris: Flammarion.

———. 2009. *The Pure Society: From Darwin to Hitler*. London: Verso.

Pico Della Mirandola, Giovanni. 1956. *Oration on the Dignity of Man*. Chicago: Henry Regnery.

Plessner, Helmuth. 1961. *Conditio humana*, in *Propyläen: Weltgeschichte*. Frankfurt-Berlin: Ullstein.

Popper, Karl. 1976. *Unended Quest: An Intellectual Autobiography*. Glasgow: Fontana-Collins.

———. 1982. *Postscript to the Logic of Scientific Discovery*. London: Routledge.

Portmann, Adolf. 1951. *Biologische Fragmente zu einer Lehre vom Menschen*. Basel: Schwabe.

———. 1965. *Aufbruch der Lebensforschung*. Frankfurt a.M.: Suhrkamp.

Ratzel, Friedrich. 1923. *Politische Geographie*. Munich: Oldenburg.

Richards, Robert J. 1992. *Meaning of Evolution: The Morphological Construction and Ideological Reconstruction of Darwin's Theory*. Chicago: University of Chicago Press.

———. 2002. *The Romantic Conception of Life: Science and Philosophy in the Age of Goethe*. Chicago: University of Chicago Press.

———. 2003. "Darwin on Mind, Morals, and Emotions," in *The Cambridge Companion to Darwin*, ed. Jonathan Hodge and Gregory Radick. Cambridge: Cambridge University Press.

Rodotà, Stefano. 2006. *La vita e le regole: Tra diritto e non diritto*. Milan: Feltrinelli.

Roger, Jacques. 1963. *Les sciences de la vie dans la pensée française du XVIII siècle. La génération des animaux de Descartes à l'Encyclopédie*. Paris: Albin Michel.

Rose, Stephen, Richard Lewontin, and Leon Kamin. 1983. *Not in Our Genes*. New York: Pantheon.

Rosenberg, Alex. 2003. "Darwinism in Moral Philosophy and Social Theory," in *The Cambridge Companion to Darwin*, ed. Jonathan Hodge and Gregory Radick. Cambridge: Cambridge University Press.

Rousseau, Jean-Jacques. 2002. *The Social Contract* and *The First and Second Discourses*. New Haven, Conn.: Yale University Press.

Sartre, Jean-Paul. 2007. *Existentialism is a Humanism*. New Haven, Conn.: Yale University Press.

Schelling, Friedrich Wilhelm Joseph. 1927a. *Von der Weltseele, eine Hypothese der höheren Physik zur Erklärung des allgemcines Organismus*, in *Schellings Werke*, vol. 1. Munich: Beck und Oldenbourg.

———. 1927b. *Erster Entwurf eines Systmes der Naturphilosophie*, in *Schellings Werke*, vol. 1. Munich: Beck und Oldenbourg.

———. 1978. *System of Transcendental Idealism*. Charlottesville: University Press of Virginia, 1978.

———. 1980a "Of the I as Principle of Philosophy, or On the Unconditional in Human Knowledge," in *The Unconditional in Human Knowledge: Four Early Essays (1794–1796)*. London: Associated University Presses.

———. 1980b. "New Deduction of Natural Right," in *The Unconditional in Human Knowledge. Four Early Essays (1794–1796)*. London: Associated University Presses.

———. 1988a. *Ideas for a Philosophy of Nature as Introduction to the Study of this Science*. New York: Cambridge University Press.

———. 1988b. *Allgemeine Übersicht der neuesten philosophischen Literatur*, in *F.W.J. Schellings historish-critische Ausgabe*, vol. IV, Stuttgart: Frommann-Holzboog.

———. 2002. *Fernere Darstellungen aus dem System der Philosophie*; Italian translation "Ulteriori esposizioni tratte dal sistema di filosofia," in *Filosofia della natura e dell'identità. Scritti del 1802*, Milan: Guerini.

———. 2004. *First Outline of a System of the Philosophy of Nature*. Albany: State University of New York Press.

———. 2006. *Philosophical Investigations into the Essence of Human Freedom*. Albany: State University of New York Press.

Schlanger, Judith. 1971. *Les métaphores de l'organisme*. Paris: Vrin, 1971. [reprint Paris, L'Harmattan, 1995.]

Schoener, Thomas W. 1989. "The Ecological Niche," in *Ecological Concepts*, ed. John M. Cherrett. London: Blackwell

Schopenhauer, Arthur. 1896. *The World as Will and Idea*. London: Kegan Paul.

Semerari, Giuseppe. 1995. "Introduzione," in Friedrich J. W. Schelling, *Lettere filosofiche su dommatismo e criticismo,* Rome-Bari: Laterza.

Sfez, Lucien. 1995. *La santé parfaite. Critique d'une nouvelle utopie.* Paris: Seuil.

Simmel, Georg. 2010. *The View of Life: Four Metaphysical Essays, with Journal Aphorisms.* Chicago: University of Chicago Press.

Simpson, George G. 1944. *Tempo and Mode in Evolution.* New York: Columbia University Press.

Sloan, Phillip R. 1986. "Darwin, Vital Matter, and the Transformism of Species," *Journal of the History of Biology* 19.

———. 1992. "Introductory Essay: On the Edge of Evolution," in Richard Owen, *The Hunterian Lectures in Comparative Anatomy. May-June 1837.* London: Natural History Museum Publications.

Smith, Christopher U. M. 1999. "Coleridge's 'Theory of Life,'" *Journal of the History of Biology* 32.

Sontag, Susan. 1977. *Illness as Metaphor.* New York: Farrar, Strauss & Giroux.

Tarizzo, Davide. 2003. *Il pensiero libero: La filosofia francese dopo lo strutturalismo.* Milan: Cortina.

———. 2004. *Homo insipiens: La filosofia e la sfida dell'idiozia.* Milan: Franco Angeli.

———. 2007. *Giochi di potere: Sulla paranoia politica.* Rome-Bari: Laterza.

Taylor, Charles. 1989. *Sources of the Self: The Making of Modern Identity.* Cambridge, Mass.: Harvard University Press.

———. 1992. *Multiculturalism and the Politics of Recognition.* Princeton, N.J.: Princeton University Press.

Traverso, Enzo. 2003. *The Origins of Nazi Violence.* New York: New Press.

Van Inwagen, Peter. 1994. "Doubts about Darwinism," in *Darwinism: Science or Philosophy?* ed. Jon Buell and Virginia Hearn. Richardson, Tex.: Foundation for Thought and Ethics.

Weikart, Richard. 2004. *From Darwin to Hitler: Evolutionary Ethics, Eugenics, and Racism in Germany.* New York: Palgrave.

West-Eberhard, Mary Jane. 1992. "Adaptation: Current Usages," in *Keywords in Evolutionary Biology,* ed. Evelyn Fox Keller and Elisabeth A. Lloyd.. Cambridge, Mass.: Harvard University Press.

Williams, Bernard. 1985. *Ethics and the Limits of Philosophy*. Cambridge, Mass.: Harvard University Press.

———. 1993. *Shame and Necessity*. Berkeley and Los Angeles: University of California Press.

———. 2002. *Truth and Truthfulness: An Essay in Genealogy*. Princeton, N.J.: Princeton University Press.

Williams, George C. 1966. *Adaptation and Natural Selection: A Critique of Some Current Evolutionary Thought*. Princeton, N.J.: Princeton University Press.

———. 1992. *Natural Selection: Domains, Level, and Application*. New York: Oxford University Press.

Wilson, Edward O. 1978. *On Human Nature*. Cambridge, Mass.: Harvard University Press.

Wright, Sewall. 1931. "Evolution in Mendelian Populations," *Genetics* 16.

Wynne-Edwards, Vero Copner. 1962. *Animal Dispersion in Relation to Social Behaviour*. New York: Hafner.

INDEX

(continued from page ii)

23 *Vampyroteuthis Infernalis: A Treatise, with a Report by the Institut Scientifique de Recherche Paranaturaliste*
Vilém Flusser and Louis Bec

22 *Body Drift: Butler, Hayles, Haraway*
Arthur Kroker

21 *HumAnimal: Race, Law, Language*
Kalpana Rahita Seshadri

20 *Alien Phenomenology, or What It's Like to Be a Thing*
Ian Bogost

19 *CIFERAE: A Bestiary in Five Fingers*
Tom Tyler

18 *Improper Life: Technology and Biopolitics from Heidegger to Agamben*
Timothy C. Campbell

17 *Surface Encounters: Thinking with Animals and Art*
Ron Broglio

16 *Against Ecological Sovereignty: Ethics, Biopolitics, and Saving the Natural World*
Mick Smith

15 *Animal Stories: Narrating across Species Lines*
Susan McHugh

14 *Human Error: Species-Being and Media Machines*
Dominic Pettman

13 *Junkware*
Thierry Bardini

12 *A Foray into the Worlds of Animals and Humans,* with *A Theory of Meaning*
Jakob von Uexküll

11 *Insect Media: An Archaeology of Animals and Technology*
Jussi Parikka

10 *Cosmopolitics II*
Isabelle Stengers

9 *Cosmopolitics I*
Isabelle Stengers

8 *What Is Posthumanism?*
Cary Wolfe

7 *Political Affect: Connecting the Social and the Somatic*
John Protevi

6 *Animal Capital: Rendering Life in Biopolitical Times*
Nicole Shukin

5 *Dorsality: Thinking Back through Technology and Politics*
David Wills

4 *Bíos: Biopolitics and Philosophy*
Roberto Esposito

3 *When Species Meet*
Donna J. Haraway

2 *The Poetics of DNA*
Judith Roof

1 *The Parasite*
Michel Serres

DAVIDE TARIZZO teaches moral philosophy at the University of Salerno. He has edited Italian editions of texts by Sigmund Freud, Gilles Deleuze, Ernesto Laclau, Jean-Luc Nancy, Stanley Cavell, Hannah Arendt, and Alain Badiou. Among his recent essays are *Homo insipiens: La filosofia e la sfida dell'idiozia, Giochi di potere: Sulla paranoia politica,* and *Introduzione a Lacan.*

MARK WILLIAM EPSTEIN has translated Lars-Henrik Olsen's *Tracks and Signs of the Animals and Birds of Britain and Europe,* Luca Peliti's *Statistical Mechanics in a Nutshell,* and Silvio Pons and Robert Service, editors, *Dictionary of Twentieth-Century Communism.* He is a co-editor of *Creative Interventions: The Role of the Intellectual in Contemporary Italian Culture* and has published on Italian literature, criticism and thought, and cinema.